frugavore

frugavore

How to grow organic, buy local, waste nothing, and eat well

Arabella Forge

with illustrations by Genna Campton

Skyhorse Publishing

Skyhorse Publishing books may be purchased in bulk at special discounts
for sales promotion, corporate gifts, fund-raising, or educational purposes.
Special editions can also be created to specifications. For details, contact the
Special Sales Department, Skyhorse Publishing, 307 West 36th Street, 11th
Floor, New York, NY 10018 or info@skyhorsepublishing.com.

Skyhorse® and Skyhorse Publishing® are registered trademarks of Skyhorse
Publishing, Inc.®, a Delaware corporation.

www.skyhorsepublishing.com

10 9 8 7 6 5 4 3 2 1

Library of Congress Cataloging-in-Publication Data is available on file.
ISBN: 978-1-61608-408-0

Printed in China

Contents

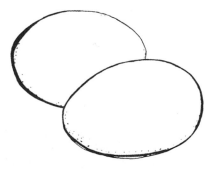

INTRODUCTION

I STARTED WRITING THIS BOOK AFTER I found myself trying to juggle two seemingly opposing things: I wanted to provide good, nutritious food for myself and my family, while also watching my dollars when I went to the supermarket.

At first, like many people today, I took steps toward healthy eating by shopping at organic food stores and occasionally at the local farmers' market. Where possible, I tried to buy organic meat, fresh seafood, and good quality fruit and vegetables. Living in a busy household, I was meticulous about throwing things out when they got moldy or stale. This made life expensive. And, looking back on it, more than a little wasteful.

Eventually, though, I started to feel frustrated. I was fed up with the high prices at small organic stores, and by the poor quality of the produce at my local supermarket. I started to look for other ways to access good food, and I began to live and cook a lot more frugally. I still wanted to buy the most nutritious food I could, and to enjoy good quality meat, fish, and vegetables. But I learned to shop more wisely, to make the most of what I bought, to waste less, and to connect more closely to where my food came from.

Along the way, more than a few changes took place at our house. The front lawn was taken up and replaced with a veggie patch (much to the horror of anyone who liked to kick a soccer ball!). We got two little hens and converted an old warehouse container into a henhouse. I also bought a second-hand freezer and started to buy food in bulk when it was in season. As well as saving us money, this made my life much simpler and easier: I always had frozen produce on hand if I needed to whip up dinner in a hurry during the week, and I could easily source fresh vegetables and herbs from the garden.

I also started to ask more questions about the food I bought. I quizzed my local butcher about the cheapest cuts of meat. I scoured the supermarket shelves for low-cost, easy to prepare foods like lentils, chickpeas, and legumes. As I experimented in the kitchen and used my family as guinea pigs, I also read more and more about traditional cooking techniques and peasant-style cuisines. Somewhere along the way I fell upon a word that I soon became very fond of. It was the magic word *frugal*, meaning to make the most of what we already have and also, wherever possible, to use *less*. People living off the land have employed frugal cooking and harvesting techniques for millennia. They did so not just to save costs and eliminate waste but also as a means of staying healthy. As I explored new ways of buying and preparing food, this notion of frugality made more and more sense.

As a nutritionist, I was lucky enough to have contacts in the food world, who were a great source of help and advice. Of all of these relationships, by far the most valuable were those with local farmers. From the farm, I could buy meat in bulk, fresh milk, and a wonderful array of other produce. I learned how to use all the thrifty bits—pork belly could be cured into bacon, chicken feet could be used to make stock for soup, and extra milk could be turned into homemade cheese or yogurt. By visiting local farms, I also gained a much better understanding of how my food was produced and what was fresh and in season.

In many respects, this approach to food represents a return to traditional peasant ways of eating. Only a few generations ago, almost everyone with a patch of land grew their own vegetables, kept a few chickens, and preserved their own fruit. What you couldn't produce yourself, you could usually find nearby: a neighbor with a lemon tree, a friend with a fresh catch of fish, a local baker selling freshly made loaves. Your diet would reflect what was locally and seasonally abundant, and you would have a clear understanding of where the food on your plate came from. Being a frugavore is all about rediscovering these peasant habits of frugality, returning to simple, fresh, seasonal food and reconnecting with the farm.

It's not just a matter of nostalgia, though: the time is ripe for a more frugal approach to food. We live in an age of profligacy. Never before have we had so much food available, and never before have we wasted so much. Celebrity chefs entertain us with elaborate meals and trendy ingredients, but most of us don't eat the sort of food we see on TV. We buy our food ready-made from the supermarket or as take-out, carry it home in plastic containers, and zap it in the microwave (doing away with any nutrients that managed to survive that far). Most of us have got into the habit of shopping only at supermarkets, and many people don't know how to prepare meals at home from scratch, or don't think they have time. When food is available immediately it's easy to take it for granted; it's no wonder that around one third of the food we buy is wasted every year.

Many people would like to eat differently but aren't sure where to start or don't think they can afford it. There can be a big price discrepancy between quality, chemical-free produce and conventional supermarket fodder. A leg of grass-fed lamb from the local organic butchery is going to cost you a lot more than the no-name equivalent from the supermarket. There are cheaper ways to access good-quality, organic produce, but very few of us know the way to our local farm or how to grow our own vegetables. That's where this book comes in: as a frugavore, you'll be equipped with the skills and knowledge to enjoy the best produce at a much better price. So instead of buying lean, organic chicken breasts at your local organic store, go buy the whole bird (and I mean the whole bird—head and feet included). Use the meat in a roast, keep the leftovers for school sandwiches, and make a good hearty soup with all the bones. You now have three meals instead of one— and if you shop at the same places I do, you'll probably find that a whole chicken costs about the same as those two skinless fillets. What's more, as you will learn on the following pages, all those extra bits—the bones, the heads, and the skin—are far better for you than the cardboard-flavored chicken breasts you might be accustomed to.

Of course, traditional peasant groups had the advantage of living close to the land; reconnecting with the farm can seem a daunting prospect for those of us living in modern cities and working long hours. But times are a-changing; more people are demanding healthier, tastier, more sustainable food, and the inner-city food market has had to adapt. In an attempt to rebuild the connection between farm and table, modern-day urban peasants are staging a food revolution. They are joining co-ops or buying clubs, or driving out to nearby farms to collect fresh food. Suburban and inner-city gardens are a great untapped resource, and more and more communities are finding ways to grow fresh vegetables on public and private land. Our food culture is changing as people seek out creative ways to improve their diets. This book will help you to be a part of this, whether you live on a farm, in a high-rise apartment, or on a suburban block.

How to spot a frugavore

So what is a frugavore? A frugavore makes the most of what they have, supports best practices in farming, wastes nothing, and grows their own food when they can. As a frugavore, you may find yourself:

- Sourcing food locally and seasonally;
- Buying food in bulk, be it meat direct from the farm or grains, vegetables, and fruits from co-ops, markets, or other sources;
- Stocking your pantry wisely with staple items that can be used as a basis for simple, healthy meals;
- Learning to cook as a peasant would, using frugal ingredients such as lentils, legumes, offcuts, cheaper varieties of meat and fish, and seasonal fruits and vegetables;
- Letting nothing go to waste—using scraps for a compost heap or worm farm, making stock with old vegetables and bones, and recycling glass jars for storage and preserves; and
- Connecting to local, grassroots food movements and exploring food resources in your local community.

How will you feel after all of this? You may notice that you have no room in your garden, as it has been taken over by pots of herbs and climbing tomatoes. Your kitchen cupboards will be overflowing with spare jam jars. You'll start contemplating replacing your coffee table with a small second-hand freezer, and you may catch yourself going to work with dirt underneath your fingernails. Don't worry: it's not a bad thing! It's all part of becoming a frugavore.

Frugavore nutrition

There's no point spending money on food that's not going to do you any good; "cheap" food is hardly good value if it's bad for you. Instead, if you want to eat and live frugally, buy the best quality produce possible and make the most of it. By "best quality," I don't mean the finest French cheeses or the most expensive bottled water. Quality food is food that is full of nutrients, grown locally, and prepared fresh at home.

Technically speaking, the term "nutrient-density" refers to the quantity of nutrients in a food compared to the number of calories or kilojoules it contains. A nutrient-dense diet will promote health and stamina and help your body to ward off disease.

Nutrient-density is also fundamental to the most important aspect of our eating pleasure—taste. I used to wonder why home-grown fruit and vegetables tasted so good compared to commercially grown supermarket produce. I've seen people swoon over the first apple of autumn, or over ripe asparagus sprouting from the ground in spring, but never over a tired navel orange, puffed up and yellowing, that has been in transit for weeks before reaching the supermarket shelf. Nature cleverly organizes itself so that when food is at its ripest,

it is also at its most nutritious peak. When food is ripe and rich in nutrients, it will be alive-tasting and bubbling with flavor and bite. This may explain why we are tempted to pick fruit from over a neighbor's fence when a tree is dripping with plums, ripe and ready to be eaten, but don't feel similarly tempted by something that has been sitting in a fruit bowl for days.

HOME-COOKED MEALS

The best way to ensure you're eating nutrient-dense food is to cook your meals at home, using raw ingredients, rather than relying on processed or pre-packaged foods. Seek out foods that have come from rich and fertile soils, and try to grow some of your produce yourself or connect to a local farm or community garden.

You don't need to be a arugula scientist to know that home-cooking is better for you than processed or ready-made meals. Processed food is treated to ensure a long shelf life and a neat appearance. The food industry uses methods such as canning, pasteurizing, refining, and irradiating to create products that can sit on supermarket shelves for months, sometimes years. Additives and preservatives are added to ensure that your food won't crumble, age, or lose its color and shape before it reaches you. Unfortunately, these methods also strip nutrients from food. That's how they work: by removing nutrients, processing eliminates the food source for micro-organizms and insects, and so prevents the product from deteriorating on the shelf.

The food industry is well aware of what is taken out of food when it is processed. That's why companies "fortify" foods with missing vitamins and minerals such as fiber, B vitamins, and iron. When a food has been broken down and stripped of its nutrients, however, it cannot be restored simply by adding nutrients later on. Nutrients require special enzymes and co-enzymes for proper assimilation. These natural combinations are only found in whole foods; they can't be recreated in a factory.

This isn't to say that all processed foods are unhealthy. Traditional food preservation methods maintained or even enhanced the nutrients in foods. Salting and air-drying can be used to preserve meat for long periods as delicious salami and bacon. Dairy products can be turned into yogurt, kefir, and cheese, which are rich in healthy bacteria and can be even more nutritious than fresh milk. Adding salt to cabbage produces sauerkraut, an excellent source of healthy bacteria and vitamin C. Foods preserved using these traditional methods can be delicious and highly nutritious (in fact, it was the addition of sauerkraut to the diet of Captain Cook's crew that made his voyage so successful). To ensure you're getting the most nutrition for your buck, however, steer clear of foods processed using artificial additives and preservatives. They may seem cheap and convenient, but you'll be missing out on the best bits—the nutrients and the flavor.

Straight from the Farm

When it comes to fresh produce, nutrient-density can be traced back to the farm. What we feed growing plants, and what they encounter in the ground, will affect the nutrients in the produce that we eventually eat. If the soil is biodynamic or organic, it will contain slow-release fertilizers in the form of manure, compost, and some minerals, providing the plant with a range of nutrients and creating robust root systems. Biodynamic and organic produce can take longer to grow than industrially farmed vegetables, but this extra time allows their roots to develop and run deep. The plant gathers extra nutrients and we end up with a healthier final product. And because plants that are rich in nutrients have a natural immunity and can repel pests by themselves, they can be grown without chemical pesticides.

Over the past fifty years, agricultural companies have worked to develop larger crops, greater yields, and hardier species. This has been achieved through selective breeding methods and high-yield fertilizers. Today's fruits and vegetables can be dropped, bounced, and stacked on the supermarket shelf without damage; they can be stored for months without any visible change. Fresh produce is cheaper and easier to access than it was fifty years ago, especially in city areas.

But greater yield and bulk have come at a cost: there are now fewer nutrients in our food. Plants are pushed to grow bigger by accumulating more water and starch, but there is not a corresponding increase in nutrients. Vegetables tested in 1980 contained significantly lower levels of calcium, magnesium, copper, and sodium than vegetables tested in 1930. As Michael Pollan has observed, "you now have to eat three apples to get the same amount of iron as you would have gotten from a single 1940 apple." We might have more food in the twenty-first century, but it is worth less in terms of nutrient-density.

With the growth of large-scale commercial agriculture, we have also lost many traditional varieties of fruit and vegetable. In our endless quest for the reddest tomato or the hardiest pear, much diversity has been lost. There were once thousands of varieties of tomato available, yet you'll be lucky to find more than two or three in most modern supermarkets. Carrots once came in white, purple, red, and yellow. They could be small, tall, stumpy, or slim. Somewhere along the line, we decided that the orange Dutch variety was the best to sell. This variety became universally available, while the others dropped out of the food chain. Many of these traditional varieties (often called "heritage" or "heirloom") are now experiencing a rebirth, helped along by local food communities, small farms, and home gardeners. They may look small and mis-shapen next to the uniform produce at the supermarket, but these heritage vegetables are packed full of flavor and nutrients. If you grow them yourself, you'll know exactly what fertilizers and chemicals have gone into them. And you'll get to enjoy the novel pleasure of discovering different colors, shapes,

and flavors. A home-grown carrot with two legs, baked with a little goose fat and fresh herbs; a black tomato, speckled with tiny red spots, drizzled with some fresh olive oil; violet gnocchi made with "purple congo" potatoes. We miss out on this diversity if we depend on uniform supermarket produce. Luckily, it's much easier than you might think to reclaim it.

The animal products we eat, whether dairy, eggs, or meat, are also affected by the nutrients in the soil and the plants we feed them. Traditionally, all livestock were grazed on fresh grass, with only minimal supplementary feeding during the drier months. Poultry, being omnivores, pecked at mixed pastures and enjoyed insects, grubs, and a small amount of grain. Their lives included plenty of exercise and sunlight.

In the push to achieve cheaper food production during the twentieth century, more concentrated feeding practices were developed. Commercially farmed livestock and poultry are now often fed a diet exclusively of grains, corn, or soy meal. They are kept in close confinement, without access to sunshine or space. Poultry, who have a strong natural instinct to run around, play in the dirt, and establish their own pecking order, are kept in small cages, have their beaks clipped, and are fed an unnatural (and monotonous) diet.

Nutritional tests have shown that animals raised this way are less healthy than their free-range, natural-living counterparts. Because they live in cramped conditions and eat an unnatural diet, they are prone to infection and more likely to need antibiotics, which end up in the products we eat. Meat, milk, and eggs from grass-fed and free-range animals, meanwhile, are rich in healthy fatty acids including Omega 3 and conjugated linoleic acid (CLA)—both healthy fats known to benefit heart health and aid in the prevention of inflammatory and autoimmune disorders. Grass-fed and free-range products also contain higher levels of antioxidants and important fat-soluble vitamins such as A, D, and E. Grass-fed livestock do not suffer from many

of the afflictions faced by their grain-fed counterparts such as acidosis, rumenitis, liver abscesses, and bloat, and they are at less risk of E-coli contamination.

The surest way to find the most nutritious produce is to reconnect with local farmers. As you'll discover later in this book, there are all sorts of ways to do this. Whether you buy directly from a farm, at a farmers' market, or through a co-op or buyers' club, a clearer understanding of where your food came from and how it was grown will help you to choose the healthiest fresh ingredients.

Is good food more expensive?

The health benefits of good food are clear. But people are always telling me that it costs too much to eat well. They quip that buying processed foods or take-out is cheaper than cooking at home, and complain that organic produce is overpriced and overrated.

There is some truth in this. The great price difference between organic or biodynamic and conventional produce is undeniable. This morning, I bought a bunch of celery at my local supermarket for around a quarter of the price I would pay for the biodynamic equivalent. Grass-fed organic meat costs more than conventional meat; a 6-pound leg of lamb at my local gourmet butcher is usually 30 to 40 percent more expensive than at the supermarket.

So if you ask me whether it costs more to buy the best quality produce through conventional retail outlets, my answer, in short, is yes it does. But there are shortcuts and backstreets you can take to access good food without going broke. By tapping into unconventional food resources and being frugal with what you buy, you can stick to a low budget while enjoying quality produce; that is what being a frugavore is all about.

For example, when it comes to my celery and grass-fed lamb, you could consider growing some of your own celery. A packet of six seedlings retails for around the cost of a bag of potato chips, and will last you for most of the season. These can be grown in pots, or in a small patch of your garden.

Similarly, to obtain grass-fed organic meat at a better price, you could consider purchasing your meat in bulk, directly from the farm, or opting for the cheaper cuts—such as casserole, mince, or offal—and exploring different ways to cook them. Every section of this book details creative ways to make the most of what you purchase, and to access good food at a reasonable price.

While I was writing this book, I wanted to test the premise myself, to find out if it was in fact easier and cheaper to eat out than to cook at home. So I conducted a little experiment. On a Monday night, I drove to my sister's place, where she lives with her husband and three young kids. I picked them up and took them out for a fast-food dinner.

Ordering three kids' servings, plus two adult meals, our bill came to around thirty-five dollars. It was a fun night out; it took no more than half an hour to drive to the restaurant and back, and we avoided all the dirty dishes. But from a nutritional perspective, I couldn't quite believe what we were eating. It was rich in salt, sugars, trans-fats, preservatives, and flavorings. It was low in nutrients, and high in calories.

The following Monday night, I thought we'd try something different. I live three blocks from my sister's place, so I invited them round for dinner. Because I didn't have a lot of time after work, I made lentil soup. When you're in a hurry, this is a pretty simple exercise: I grabbed a packet of lentils from the pantry, a few bay leaves from the garden, some stock from the freezer, and some carrots and celery from the fridge. It took me ten minutes to get the ingredients ready and forty minutes to cook. While the soup was cooking, I picked some arugula from a pot outside my kitchen door and threw together a salad, and then put a sago and coconut pudding on to cook for dessert. The entire preparation took me half an hour. The total cooking time was about an hour. The meal, all ingredients included, cost less than half of what we paid

for our meal at the fast-food outlet—plus, there were leftovers.

As we ate, we chatted about the concepts of price and convenience. I calculated that of the recipes I cook on an average weeknight, nearly all are cheaper and more convenient than eating out—cheaper even than the cheapest of drive-thrus. People are always telling me that they choose to eat fast or processed food because it is cheaper, but this is just not true! By being frugal with food—making the most of what you have, growing some herbs or veggies yourself, and connecting to local food resources—you can save money on groceries, and spend less time shopping or picking up ready-made meals. Don't believe me? Well, you'd better keep reading then.

EAT AS A PEASANT
WOULD EAT

"The old peasant kitchen habits of frugality [involved] making stock out of bones, pickling and salting in times of glut, stocking the pantry, using diet to care for the sick and the elderly, making good food out of few and simple ingredients."

—Elisabeth Luard, *European Peasant Cookery*

Peasant-style cooking embodies all that it is to be a frugavore: choosing nutrient-dense foods from local sources, making do with what is available at different times of the year, and being as self-sufficient as possible.

More and more, we are realizing that we have a lot to learn from peasant diets. The food we eat today is not nourishing us as it should. We are largely unaware of where it comes from, but we are beginning to understand that many modern food-production methods come at a great cost to the environment.

Some might see the word *peasant* as a derogatory term: maybe a starving farmer from the Middle Ages who cannot read or write and has too many children to feed with too little food. In fact, the word is derived from the fifteenth-century French word *païsant*, meaning a person from the local *pays*, or countryside. By definition a peasant is any person who lives or works close to the land.

The peasant diet has been described as "simple and nourishing," and their health was all the better for it. When they weren't affected by wars or food shortages, peasants lived surprisingly healthy lives. Elisabeth Luard has described peasant communities in Spain in the 1920s as "communal, supportive and hardworking," and Dr. Weston A. Price, in his travels around the globe during the 1930s, found them to display extraordinary health, strength, and vigour.

Existing exclusively on a traditional diet drawing only on local resources, peasants relied on their food to keep them strong and protect them from disease. Their food preparation methods were well considered and thrifty, designed to get as much value and nourishment as possible from every dish.

Most importantly, nothing was ever wasted. Food had to be harvested with the utmost diligence during times of scarcity, and stored with the utmost care during times of abundance. Peasants were thrifty and frugal with food because they had to be; their diets were dictated by limitations such as the fertility of the soil, the seasons, the weather, and their region's habitat.

So we have a lot to learn from peasant cultures, and food choices are just the beginning. Peasants lived a lifestyle that was supremely self-sufficient. They made the most of what they had on-hand and wasted nothing. All waste produced on the farm was used for some other purpose: food scraps went to the chickens or were used as compost; chicken poop fertilized the plants. Only the bare minimum was purchased or traded and nothing was carelessly discarded.

Today, when it comes to food, we are inundated with choices. At the supermarket we find foods out of season, imported from other countries, fresh, frozen, processed, pasteurized, and irradiated. Meals can be bought ready-made and reheated in seconds with the click of a button. Food might be more readily and abundantly available, but it is worth less in nutritional terms and lacks the flavor of locally grown, homemade fare.

It's no surprise that many people now yearn for a simple and frugal diet, and for the freshness and flavor of plain old peasant cooking: fresh bread with lashings of butter and homemade jam; a hock of bacon simmering in a soup of split peas and spices; tender tomatoes bursting with flavor at the beginning of the season.

Peasant fare does not need to revolve around plain or bland ingredients. In fact, traditional peasant foods were full of richness, flavor, and diversity. European peasants living near the ocean enjoyed such scrumptious foods as salmon, oysters, and crayfish, while those living inland had slow-cooked snails, raw-milk cheeses, superb truffles, and foie gras. Even the simplest of dishes tasted deliciously good. Unfortunately for the food industry, this cannot be recreated with additives, preservatives, or modern processing methods. The joy of this sort of food lies in its freshness, and in knowing that what you are eating was made by you, your family, or friends, or someone in your local neighborhood.

Not surprisingly, as world leaders in gastronomy, the French have always valued their traditional peasant cooking as much as they do their *cuisine bourgeoise*. Curnonsky, editor of *Larousse Gastronomique* and founder of the Michelin Star rating system, believed that regional peasant-style cooking possessed some of the richest of gastronomic treasures, and that French cooking should always strive to respect "the taste of things as they are." The French have gone to great lengths to preserve their traditional food and farming culture, and it is for this reason that they've been able to maintain such a flavorsome, nutritious cuisine.

So, what did peasants eat? Most foods were bought fresh from their source and any processing, cooking, or preserving was done in the home or communal kitchen. As much food as possible was grown at home: no garden space was wasted. All parts of the animal were used, and any animal or vegetable scraps were put back into the land as compost.

One-pot meals such as soups and stews were common, as they required only a single heat source and were a good way to use up tougher cuts of meat. Traditional diets included a range of bread and other wholegrain products, which were prepared using traditional methods to render them optimally nutritious and easy to digest. Dairy products were consumed both raw and fermented in the form of yogurt, butter, kefir, and cheese. Meat and other animal products came from both wild and domesticated animals. Livestock and poultry were grass-fed and free-range; they were not fed large quantities of grains or corn, nor any soy meal.

Peasant food also depended on seasonal availability. Food that was in season would feature as the main aspect of a meal: the latest garden peas, fresh cheeses locally made, or a new batch of eggs from the chickens. The season's ingredients would be cooked from scratch or preserved for later use using traditional methods. Place was everything, and each local area had its own specialties.

As they say in France, "The triumph of Marseille, it is only good when eaten in Marseille. Don't try to eat it in Paris." Wherever you were, great respect was shown for the food on the table. The traditional prayer before a meal demonstrated gratitude for the food, and for the farmers and animals who produced it.

It is also telling to consider what peasants *didn't* eat. Refined sugars, high-fructose corn syrup, and artificial sweeteners did not exist. Natural sweeteners such as honey, molasses, malt, and maple syrup were used when they were available. Seed oils like cottonseed, safflower, and sunflower oil were burned for light and used in traditional remedies, but were not generally used for cooking. Instead, animal fats in the form of eggs, cream, butter, and oil were highly prized, as was cold-pressed olive oil. Hydrogenated fats such as margarine did not yet exist. The only soy foods consumed were fermented: soy sauce, tempeh, miso, nato, and tofu. Soy "milk," soy-based imitation foods, and soy additives were not known.

Needless to say, genetically modified foods did not feature in traditional diets.

Waste not, want not

Peasant frugality was not confined to the kitchen. Out of necessity, traditional cultures kept waste to a minimum and squeezed as much use as possible out of what they had. Even a generation or two ago, this attitude was common. Growing up, I was constantly reminded that food should be used frugally and enjoyed to the last morsel. I remember my mother removing a soured pumpkin from the compost heap (after I had sneakily thrown it there), trimming off the moldy edges, and using it for dinner. It was surprisingly enjoyable. She always kept an eye out at the supermarket for food that was on sale because it was nearing its use-by date, and all our leftovers went to our backyard chickens or onto the compost heap. For my parents' generation, these habits were fairly standard. Growing up after the war, they and their parents appreciated the energy required to produce a loaf of bread, a stick of celery, or a can of beans.

Today, we have forgotten how to be frugal. If electrical goods break, it is often cheaper to buy new ones than to have them repaired. We buy food in take-out containers and disposable plastic bags without a thought. We buy and discard food carelessly, knowing there will always be more for us to purchase tomorrow. Along the way, we have forgotten many of the clever tricks used by earlier generations to reduce waste.

But with rising food prices and overflowing landfills, we are seeing a quiet backlash against this profligacy. "Dumpster divers" scour garbage skips for food that would otherwise go to waste. I haven't dumpster-dived myself, and I can't say I want to, but I can appreciate what they do and why they do it. In Australia, food waste is estimated to be 30 million tons annually.

During the past decade, we have become increasingly environmentally aware; we are conscious of our carbon footprint and of the need to save energy and use fewer natural resources. But we also need to generate less waste. Our landfills are constantly growing. In Victoria, Australia, where I live, food and green waste account for almost half of the municipal waste we send to landfills. This type of waste produces methane, one of the worst greenhouse gases. Obviously, recycling is an integral part of reducing the impact of landfills, but recycling also requires energy, and only a proportion of what we put in our recycling bins ends up being properly reused. To actually make a difference, we need to use less and be frugal with what we purchase.

Happily, in a kitchen environment, this is not as hard as it sounds, and this is where those rediscovered peasant habits come in handy. Sending organic waste (any plant or animal matter) to a landfill is a missed opportunity for frugavore cooks and gardeners. We could reduce our waste by almost half by recycling organic matter in our own backyards. For instance, you can take your kitchen scraps or lawn clippings and put them in your compost. Start a worm farm and use your stale vegetables and newspaper scraps as worm-food (the liquid from a worm farm is an extraordinary fertilizer and can produce stupendous results in your vegetable garden, season after season). Re-use glass jars to store homemade jam or preserved fruits and vegetables. Use tired vegetables and meat scraps to make stock. Not only will these things reduce waste, they'll also save you money by letting you squeeze much more value out of the things you've bought, grown, or made. Similarly, by buying fewer processed foods and purchasing more nutrient-dense raw ingredients, you will be using less plastic and packaging. So less waste is good for us, really and truly.

How to minimize waste in your home

- Produce as many things as you can yourself, or source them locally. Grow some of your own produce, keep chickens, and shop directly from a local farmer where possible.

- Get a compost bin, worm farm, or bokashi bucket for your kitchen waste. Use this to fertilize your garden and grow a blossoming vegetable patch.
- Consume less: only buy products that you really need. Try to avoid resorting to "retail therapy."
- Shop locally and avoid products that are processed or have a lot of plastic packaging.
- Recycle old clothing as rags for cleaning.
- Recycle old jars for jams and preserves. Use plastic containers to store leftovers and pantry staples, or for cleaning products.
- Stop using paper and plastic bags: purchase a recycled cotton, hemp, or string bag.
- Bottle your own filtered tap water instead of buying plastic bottled water.
- Practice "positive pilfering": if someone is throwing something out that could be used, grab it!

We have a lot to learn from the peasant habits of frugality—choosing delicious and nutrient-dense food, making the most of what we have, and wasting less. Obviously we can't return completely to the pre-industrialized way of life. But we can do the little things— and these little things can make a huge difference. Understand where your food comes from, choose food with less packaging, waste less, and recycle everything you can. Lastly, at the end of the day, when you sit down to enjoy your meal, take some satisfaction from knowing where your food came from. Give thanks for the person who cooked the dish, for the farmer who grew each plant and animal … and for the lucky soul who'll be doing the cleaning up.

SOURCING YOUR FOOD

"Locating better food is not something that you do once and forget. It becomes important to continually learn which fruit stand has the best items and which farm or farmette is worth a little 'drive in the country.'"

—Rex Harrill, farmer, Keedysville, Maryland

PEOPLE ARE ALWAYS TELLING ME THAT IT costs too much to eat well. I used to think the same thing. Every week I would drive to an inner-city organic food store where the fruit and vegetables were limp-looking and expensive. If I didn't feel like blowing the weekly budget, I could go to the supermarket next door, where they sold plumped-up fruit and vegetables for a quarter of the price, but with very little flavor or personality and possibly some mild pesticide residue. I knew what nutritious food was, but accessing it was another matter.

It was around this time that I started to dabble in gardening. Growing some of my own food, even in a modest way, changed how I thought about fresh produce. I found that a small patch of arugula and silverbeet could provide a fresh green accompaniment to any meal. Fresh herbs were easy (bordering on idiot-proof) to grow and added flavor to any dish. With time and a little help from some good gardening books, we were able to grow up to half of our own produce in our standard suburban backyard. This saved us money and time (no more last-minute trips to the greengrocer) and provided us with the most nutritious source of food possible. We also acquired a few chickens, happy little hens that provided an egg or two every day. Living in a suburban block, this was all we could fit, but it saved us considerable time, money, and effort in putting food on the table. We now get our food from a combination of our own backyard, a local farm, and a few organic food stores. If dinner isn't organized we can always throw together something nutritious and easy like a tasty omelet or toad-in-the-hole, garnished with a fresh green salad.

Becoming more self-sufficient made me realize how expensive and unnecessarily stressful my old approach to grocery shopping was. I used to think that I had two choices when it came to fresh food: healthier but expensive organic produce, or cheap but flavorless supermarket vegetables. Growing some of my own made me realize that tasty, nutritious food doesn't have to be expensive, and prompted me to seek out alternative sources for those things I couldn't grow myself. Most of us won't ever be completely self-sufficient. Not everyone has the time and space to have a garden, and not everyone can keep their own chickens. But there's no need to feel restricted to supermarkets and organic shops. There is an increasing demand for alternative ways to access good food. New avenues are opening up, allowing consumers to buy fresh, healthy food at much more reasonable prices. This section of the book outlines some of these different options. You might shop at a farmers' market, join a co-op or buying club, or even drive out to a local farm. You may also want to explore some of the grassroots movements described later in this chapter. They can be a great source of information about where to find local, sustainable food. You might be surprised to learn what is already happening in your area.

Some terminology

Whether you're buying directly from the farm or through a market or co-op, you'll want to know a little about the farm your food is coming from. A good farm employs traditional methods to grow nutritious, natural produce. It might use organic or biodynamic methods, or simply employ traditional techniques to avoid the need for chemicals and pesticides. Before we look at different places to buy your food, here's a guide to some of the terms you might encounter along the way.

Organic: Organic farming is based on the idea that food should be produced using natural methods and natural additives. It does not use synthetic chemicals, pesticides, or fertilizers, either on the produce itself or on the soil. It does use natural fertilizers such as animal manures, blood and bone, composts, rock phosphate, lime, gypsum, and dolomite.

Advantages of organically grown produce include:

- A reduced environmental impact. Organic farms are better at sustaining diverse ecosystems, are more energy-efficient, produce less waste, and do not consume or release synthetic pesticides into the environment.
- Better tasting, more nutritious produce. Studies of organic fruit and vegetables have shown that they have higher levels of many nutrients and antioxidants. Blind taste tests have found that organic produce often tastes better.
- Better living conditions for animals. All food fed to organically farmed animals must be certified organic. Furthermore, animals are not treated with unnecessary antibiotics, hormones, or genetically modified organizms.

Different organic certification systems exist in different countries. Keep in mind that thanks to hype surrounding organic labeling, and the higher prices farmers can charge for organically certified produce, there is the occasional flaw in the system, particularly when it comes to animal products (see the "Meat" chapter for more information). In some countries, particularly in the Uinted States, organic certification is expensive and difficult for small family farms to obtain. If you are able to connect directly with a local farmer and find out how your food was grown, this will be far better reassurance than any certification label. Personally, I'd prefer to buy my food from a local farm that grows good quality produce that I know and trust, with or without certification, than from an organic supplier in another country whose produce has been shipped around the world to get to my dinner plate.

Biodynamic: Biodynamic agriculture is an "enhanced organic" method, inspired by a series of lectures given by Dr. Rudolf Steiner in 1924. It has developed most widely and successfully in Australia. It involves a range of techniques including biodynamic soil preparations, composting, and cultivation patterns based on the lunar calendar, all of which enliven the soil with energy and nutrients. For fresh fruit, vegetables, and animal products, biodynamics is probably the best assurance of quality, as the animal or plant will have been raised in optimal conditions with fertile soils and grass-based feed. Biodynamic farming has been shown to produce more friable, fertile soils and healthier plants and animals. Biodynamic products can be expensive, however, especially if you are buying them on a regular basis. If you want to save money, try to use your produce frugally, become a biodynamic home-gardener, or join a local community co-op to share the costs.

Free-range: This certification system applies to poultry and some livestock (especially pigs). A "free-range" label indicates that the animals have lived outside a conventional feedlot and have had access to fresh air and fresh pasture, all of which are good things. However, the label does not guarantee that the animal was raised on a healthy diet. Pellet-feed, grains, and antibiotics are sometimes fed to "free-range" animals, so look for other certification systems in addition to this one if you can.

STRAIGHT FROM THE FARM

By buying directly from a local producer, you can often save money on excellent, nutrient-dense food. Most retailers bump their food prices up by 100 percent, so if you can buy directly from the farmer you can avoid many added costs. Many social and environmental benefits also spring from developing a direct relationship with producers. You will be helping local farmers to maintain and grow their farms, while reducing food miles by purchasing unprocessed local food. So much of our inner-city lifestyle depends on the quality of the produce from our farms. Similarly, the farmer's survival depends on his ability to sell his wares consistently and at a good price. If we can connect directly, the benefits are not just financial—we will also improve communication and understanding, and help to ensure the quality of our food into the future.

Farmers' markets: A good place to start is by shopping at a local farmers' market. Farmers' markets provide a direct link between the farmer and the consumer. They are a terrific source of local and seasonal produce (and remember—food that's in season will be cheaper and better for you) and a great way to meet the people who grow your food. You can develop long-term relationships with local farmers and sometimes (if your farmer is a flexible one) order produce to suit your particular needs. Farmers' markets are popping up everywhere. See the "Resources" section for some help getting started.

Farmgate sales: Another alternative is to drive to the farm itself and purchase food onsite. This can require a bit of effort and you may have to drive for several hours, depending on where you live. You could share the drive with friends and take turns making the trip, or purchase your food in bulk and freeze it. Even if you only try it once or twice, farmgate sales are a great way to see for yourself where your food comes from.

From a legal perspective, different regulations exist governing direct farm-to-consumer sales, so you'll need to find out what is legally available in your local area. In some instances, you may have to join a buyers' club, co-op, or community-supported agriculture group.

Buyers' clubs and co-operatives: Buyers' clubs and food co-ops are another way to purchase food directly from the farm or wholesaler. Members usually pay an entry fee to join, and can then purchase food from any of the co-op's participating vendors. Clubs and co-ops are often based in a warehouse or retail shop. Others have a central spot from which members collect their purchases but conduct all their transactions online. These clubs enable members to purchase food below the normal retail price while supporting small-scale vendors.

The co-ops I've visited over the years have been started by passionate, health-conscious foodies who are keen to establish a direct link with local farms or wholesalers. The range of products available through co-ops is impressive. They often sell such things as homemade sauerkraut, jam, and raw milk—typical peasant fare that was widely available only a few generations ago, but has been lost with the shift to large-scale food production and supermarkets. Members can also buy products at wholesale prices from major food suppliers (the same companies that supply large supermarkets). I recently visited a friend in Manhattan who is part of a buying club of over 900 people. They do all their ordering online and meet the farmer (an Amish man from Pennsylvania) at a pick-up location once a week. According to the members I spoke to, it's easy to take part and gives them access to fresh milk, vegetables, and meat, and homemade preserves, all direct from the farm. The prices are significantly lower than those in organic food stores, and members enjoy the chance to meet the farmer and his family. Clubs may also showcase products made by their members. My friend Becca makes batches of her famous Beccaroons—a delicious almond-flavored macaroon—in her home kitchen and supplies them to her local club.

So clubs and co-ops can give you access to wholesale produce, homemade goods, and produce direct from the farm. There is also a wonderful feel-good factor; you'll be part of a local community, and directly supporting local producers.

Community-supported agriculture: A community-supported agriculture group or CSA consists of a group of individuals who pledge support to a local farm by buying a share of its ownership. The farm then becomes the community's farm, with growers and consumers sharing the risks and benefits of food production. Most CSAs involve a weekly delivery or pick-up of vegetables and fruit, and sometimes dairy products and meat. The term CSA is more common in the United States, but similar systems are found worldwide.

To get a CSA started, you need a group of interested, like-minded friends and a willing farmer. There are several websites (see the "Resources" section) dedicated to helping people to establish CSAs in their local areas.

Community gardens: Of course, buying from a farmer isn't the only way to access good produce: you can also grow your own. A community garden is a single piece of land gardened collectively by a group of people. Gardeners usually rent a plot of land, often from the local council, on which to grow their own fruit and vegetables; sometimes there is space for small animals such as chickens or quail. This movement has been extremely popular in Australia, with over 350 community gardens in Melbourne alone, ranging from local-council establishments to "guerilla gardening," whereby people plant on nature strips and in the centers of roundabouts. I would love one day to see every extra pocket of land taken up by edible gardens, fruit trees, and climbing sugar-snap peas. Victoria could become the "edible state" rather than the "garden state," and we could take pride in our delicious self-sufficiency rather than in wide expanses of dull green lawns.

Community gardens are an excellent resource for people who don't have enough space to grow their own vegetables at home. They also foster an important sense of community among people from different cultural backgrounds, and allow people to well and truly get to know their neighbors. Other programs often spring up around community gardens. "Urban orchard" and "food swap" schemes, for example, allow people to swap produce they have grown themselves and don't need. Last week I ventured to a Melbourne community garden with an excess of lemons from my tree at home. I took them to the food-swap table and traded them for a few bulbs of home-grown garlic. I *love* garlic, so I went home very satisfied.

Land shares: Land-share agreements enable people to rent vegetable-growing land from private owners; they are an extension of the community garden concept to privately owned land. This is happening on a local level as communities develop different and innovative ways to share land between them.

One of the first land-share projects in Melbourne started close by to where I live. A woman living on a reasonably large suburban block decided to divide her backyard into garden plots and rent these out, for a minimal price, to her neighbors. She now has a back garden overflowing with fresh produce. Happy-go-lucky chickens roam around in the lap of luxury, producing delicious eggs for all the tenants.

In the UK, Hugh Fearnley-Whittingstall, founder of the well-known River Cottage, has created an online meeting spot where interested land-owners and land-renters can find each other and join forces. We need more meeting spots such as this! If you're interested in establishing a land-share agreement, start inquiring in your local area, or find an interested group of people via the internet. Alternatively, try posting notes in health-food stores, community centers or libraries, and you may be surprised by what you find.

Grassroots movements

The groups listed here are just a few of the many food-related networks developing around the world. They are a great way to meet like-minded people, share information about buying and cooking good food, and find out about foodie-related events. If you feel like you're in a good-food no-man's-land, joining one of these grassroots networks can be a good place to start.

Slow Food: The Slow Food movement began in 1986 in Italy and has since inspired a worldwide movement against "the universal folly of Fast Life," including fast food. Slow Food now has over 800 local *convivia* or chapters throughout the world. There is a membership fee—usually around sixty or seventy dollars per year—which allows you to take part in local events and meetings. Slow Food has done great things to champion traditional food production, local food economies, and biodiversity. They also host fabulous dinners, showcasing the latest seasonal produce and traditional cooking methods. These are often pricey, but are well worth it.

Localism: The local food movement encourages consumers to buy food produced in their local area. The intention is to build more self-reliant food economies and to encourage people to have a closer connection to local producers. Eating locally also reduces transport, packaging, and processing, contributing to more sustainable food production and fresher, healthier products. A "hundred-mile diet," whereby people only consume food produced within a hundred-mile radius of their homes, has become popular in the United States, Australia, and the UK. This movement has done a lot to promote local food culture and traditional indigenous cooking methods. Locavores don't just shop locally, they also cook locally, with kitchen co-ops and cooking classes designed to foster community spirit. Anyone can be a locavore—have a look online and see if there is an active network of support near you.

Local food currencies: Some communities have developed local "food currencies" for use in specific towns or regions. Local groups print their own money (for instance, Berkshire in the United States has the "Berkshare" and Lewes in the United Kingdom the "Lewes Pound"). Local vendors can choose to trade with the local currency in addition to the official national currency. As well as the feel-good factor, there is often a monetary benefit for consumers. In Berkshire, for example, shoppers using Berkshares receive a 5 percent discount. This encourages consumers to shop locally, while in theory buffering the local food economy against global economic upheavals.

The Weston A. Price Foundation: The Weston A. Price Foundation (WAPF) is a grassroots movement active in over seventeen countries. It promotes traditional, nutrient-dense foods, home-cooking methods, and the pioneering work of Dr. Weston A. Price, a dentist who studied the diets of traditional cultures in the 1930s.

WAPF members have been active in starting many of the schemes mentioned earlier in this section, such as CSAs and co-ops. Local chapters provide listings of food resources in the area and enable like-minded people to exchange information and ideas. Some branches also run cooking classes and workshops. Some WAPF information is available free online; other resources are available only to paid members. Go to www.westonaprice.org to find out more.

FEEL AT A LOOSE END?

If you feel like nothing is happening where you live, the first step is to get a group of like-minded, enthusiastic friends together. Local community groups or public noticeboards can be a good way to find one another. Once you have a few interested people together, you can start forging links with local farms or food networks. Remember, there's power in numbers! Just keep in mind that rules can vary depending on where you live, so you'll need to check your local regulations before sourcing any of the foods mentioned in this book.

*

Our relationship with food is changing. Consumers no longer have to depend on what supermarkets and organic shops choose to offer them. People are finding creative new ways to access the best quality produce straight from the farm, and at more frugal prices. In the city, more and more local governments are accommodating demands from residents who want to create community gardens, farmers' markets, or land-share arrangements. Whether you buy from a farmers' market, a co-op, a community garden, or straight from a farm gate, you'll have the satisfaction of knowing exactly where your food has come from. You'll be supporting your local economy, and saving money at the same time. You'll also, I promise you, have a lot of fun. So, what are you waiting for? Go find out what's happening near you!

THE FRUGAVORE KITCHEN

"Before enlightenment—washing dishes, carrying water;
after enlightenment—washing dishes, carrying water."

—Lao Tsu

G OOD FOOD ISN'T JUST ABOUT THE FINAL product you put on the table. It is a reflection of your entire cooking environment—your kitchen, bench tops, cooking equipment, and garden, and even the quality of the air. The kitchen is the heart of the house. How you clean it, chop up your food, and even heat your meals will not only affect the nutrients in your food, but also the flavor of every bite. So invest in some good old-fashioned cleaning products that won't spread harsh chemicals, grab a simple timber cutting board, look for some old pots and pans at the thrift store, and take your microwave to the dump. Air out your kitchen, clean out your freezer, and keep some pots of herbs by the back door. Your frugavore cooking experience is about to begin.

CLEANING

Only a few generations ago, housewives were able to keep a clean and tidy home with just a few staples: some bicarbonate of soda, vinegar, lemon juice, cleaning cloths, and good old elbow grease. We have been led to believe that anti-bacterial chemicals and bubbly soaps are necessary to get homes really clean, but this is just not true. Bacteria are everywhere around us—on our skin, in the air, on every surface that we touch. If we try to completely eliminate these bacteria from our environment, it disrupts the natural balance of nature and chemical-resistant strains can develop. Many studies are now showing that children require some exposure to bacteria to assist in their natural immune development.

What substances should you be wary of? For starters, any cleaning product labeled "danger," "hazardous," or "poison" should not be used around food. The chemicals that have been found to cause the most damage include the concentrated form of chlorine, which is found in many dishwashing detergents. This can have lasting or even fatal effects if ingested or swallowed. The lye and ammonia found in many oven cleaners can cause damage to the respiratory system and skin. Traces of these chemicals can enter the food we eat, either via residues on our plates and cutlery or through the air in the oven. The chemicals in cleaning products can also have lasting effects on the ecosystem after they've been washed down the drain. The phosphates used in many automatic dishwasher detergents, for instance, have been shown to kill fish and other organizms in our waterways.

So, what to use instead? Plenty of trendy, eco-friendly cleaning products are now popping up in supermarkets and health-food stores. These products are usually a better alternative than conventional products, but they can be expensive. The truth is you don't have to spend loads of money on expensive cleaning products. Natural, safe, inexpensive alternatives, just like our parents and grandparents relied on, are cost-effective and easy to use. Instead of having a different cleaning product for every surface, you just need a few simple basics and some good recycled cloths.

BUYING READY-MADE CLEANING PRODUCTS

- Be wary of any product labeled "danger," "warning," or "poison": you don't want anything toxic in your kitchen!
- When interpreting "eco-friendly" or "organic" claims on a label, look for the product's degree of biodegradability. The more specific, the better: "Biodegradable in five to seven days" is much more reassuring than just "natural" or "earth-loving."
- Look for "phosphate-free" and "solvent-free" wherever possible.
- Look for products that are plant-based. If you are unfamiliar with the chemicals listed on the label, either look them up or don't buy it.

- To reduce waste, re-use your cleaning bottles. Many food stores and co-ops now provide refillable containers.

MAKING YOUR OWN

Making your own cleaning products is easy, effective, healthy, and incredibly cheap. To get started, you'll need:

- Bicarbonate of soda
- Vinegar
- Borax
- Salt
- Lemons
- Eucalyptus oil
- Castile liquid soap
- A spray bottle
- Some recycled rags or old microfiber cloths

With the recipes in this section you can turn these basic ingredients into cleaning products for every surface. Always try to use the smallest amount of cleaning product possible. Even inoffensive products such as bicarbonate of soda can cause damage to the environment if used in excess. Water alone is an excellent solvent and good cleaning can be done with a strong arm and a damp cloth.

To clean bench tops, sinks, taps, and floorboards: Use a 50:50 solution of vinegar and water in a spray bottle; or a 3 percent saltwater solution (¼ tablespoon of salt for every 6 ¾ tablespoons of water); or sprinkle some baking soda on a damp cloth, wipe the surface to be cleaned, then rinse with clean water; or mix a small amount of baking soda with liquid Castile soap to get countertops extra shiny. To finish, add a few drops of lemon juice for a lovely aroma.

Casserole dishes: To remove stubborn baked-on leftovers, put 2 tablespoons of salt into the dish and fill it with boiling water. Let it stand until the water cools, then wash the dish as usual.

Dishcloths: Soak dishcloths and muslin cloths overnight in a 1:3 vinegar and hot water solution (1 cup of vinegar for every 3 cups of water). In the morning, scrub them if necessary, then rinse them. Air-dry them in the sunshine or rest them on an indoor heater.

Dishwashers: Put some bicarbonate of soda in the soap dispenser and run the wash cycle. This will clean your dishwasher, and can also be used for dishwashing instead of dishwashing powder.

Ovens: Make a solution of 4 ¼ cups of warm water, 1 ½ teaspoons of borax, and 1 ½ tablespoons of liquid soap. Spray the solution onto the oven's interior, wait twenty minutes, then clean with a rag and rinse with clean water. For tough stains, sprinkle a thick coating of bicarb soda and leave it to sit for an hour or two. Wipe off with a damp cloth.

Pots and pans: Bicarbonate of soda works well for scrubbing pots and pans, and is especially fantastic on burnt pots.

Trash cans: Splash some vinegar or lemon juice into the bottom of the can. Leave it for twenty minutes then rinse it clean. This should neutralize any bad smells.

Stainless-steel stovetops: On a damp rag, combine a dab of liquid Castile soap with a sprinkle of bicarb soda. Wipe your stovetop with the rag to remove stains and grease. Then, with a fresh damp cloth, wipe the stovetop again. Sprinkle any really hard-to-shift marks with bicarb soda, leave it for ten minutes, then scrub it off.

Wooden cutting boards: Wash the cutting board with warm soapy water, then rub it down with a handful of salt and leave it for five or ten minutes before rinsing it clean. If your cutting board starts to feel dry and brittle, rub it down with almond oil or olive oil and leave it overnight so that the oil can soak in.

KITCHEN UTENSILS

Too many cooks these days get hung up on having the latest equipment—shiny pots and pans, non-stick coatings, and the newest technology. But food doesn't need to be created with the latest appliances to look and taste good. If you want a healthy cooking environment and a nourishing finished product, stick to the basics, just like our grandparents did.

Cookware: Where possible, try to avoid cookware made with "non-stick" coating as this has been found to contain the chemical polytetrafluoroethylene (PTFE). PTFE has been associated with numerous chronic and acute health problems. If you do choose to use non-stick cookware, never heat it to high temperatures, as this appears to release more of the toxic emissions.

A better alternative is heavy stainless steel, cast-iron, stoneware, or enamel-coated pots and pans. Stainless steel or copper (not aluminum) work well for everyday steaming, frying, or boiling. If you can, it is worth investing in one or two large, heavy enamel pots. They are excellent for long, slow, gentle cooking as they retain the heat in their thick, cast-iron walls. These old-fashioned, heavy-set pots aren't always cheap, but they do last a lifetime. Many thrift shops sell good quality old pots and pans, and I've seen some good buys on eBay. My own favorite slow-roasting stew-pot was bought ten years ago from a thrift shop; it must be about forty years old. It cooks odd cuts of beef and lamb second-to-none and could easily out-do its counterparts in modern cookware departments. So choose your pots wisely—and don't worry about superficial scratches or chips; it's what's inside that counts.

Cutting boards: Wooden cutting boards that haven't been sealed or treated are the best choice for the home kitchen. Natural timber breathes easily, so repels bacteria and bugs better than plastic boards can. Clean your timber cutting board regularly and always allow it to breath when you are not using it. Keep it on the bench top or propped up against the wall, rather than tucked away in a stuffy drawer.

Microwaves: Microwaves are a convenient appliance for busy households. But microwaves do more than just reheat food quickly. They radiate heat using a form of electromagnetic energy, which destroys and deforms the molecules in food and depletes many of the important nutrients. As an alternative, choose an oven with a fan-force option. We have at home a "double oven"; one oven is 35.4 inches wide and the other is 11.85 inches wide. The smaller oven heats up quickly and uses less energy to reheat meals. With the fan-forced heat, it takes only ten or so minutes to reheat a dish—nearly as quick as a microwave, and much healthier.

Freezers: The best investment we ever made was a large freezer, which we keep in the garage or back porch. We store meat and grains that we've purchased in bulk, as well as stock and leftovers. The freezer saves me much time and worry. If I come home late at night, I can defrost something quickly and easily—some sausages, or some mincemeat for hamburgers or meatballs. I combine this with a few fresh greens from the garden and any vegetables that happen to be in the fridge, and ta-da! dinner is made. No shopping and very little planning required.

MEAL PLANNING

If you'd like to cook from scratch and have a busy job or chaotic family, you'll need to start planning. Get the pen and paper out. It sounds simple, but it always surprises me that people who are meticulously organized with their work schedule, gym routine, and weekend activities often leave food and cooking to look after themselves.

Start by choosing a day that is relatively stress-free—perhaps a Saturday afternoon or Sunday morning. Use this time to take stock of what's already in your cupboards and plan your meals for the week. Once your meals are planned, the rest is fairly straightforward. Cooking as a frugavore is not actually time-consuming, it just requires a little thinking ahead. So if you are cooking chickpea soup on Wednesday night, you need to put the chickpeas on to soak on Tuesday night or Wednesday morning. Most of the dishes in this book have been adapted to be work-friendly. If you stock your fridge and freezer wisely, you can easily whip up a tasty dish in a matter of minutes on any given weekday.

I make Saturday my market day, when I visit the local food market with my shopping list for the week. Saturday afternoon is my cooking and organizing time. Before I put away my shopping, I clear out the fridge. Any stale vegetables—floppy celery or tired carrots—can be used to make stock with the fresh bones I've purchased at the market. Anything else that looks a bit spoiled goes to the compost or the chickens (our chickens love Saturdays!). I leave the stock on to simmer for twenty-four hours and load the freezer up with it on Sunday. Having stock on hand means soups can be prepared quickly and easily throughout the week. I also use the weekend to check on the garden, reload the compost, and ensure that the hens are happy. After this, the rest of the week is fairly smooth sailing (well, from a cooking perspective, at least).

Lunch options for the nine-to-fiver

It's easy to forget about lunch until you start to feel hungry at noon and find yourself rushing out to buy an overpriced sandwich. But with a little planning, you can save heaps of money and time by packing your own. If you are short of lunch ideas, try to think beyond the obvious. You might buy a small thermos (these are usually cheap; I picked one up at a thrift store for two dollars), in which you can store hot soup or broth. Recycle plastic containers and bring in leftovers from the night before. Dishes such as casserole and baked beans taste better if they are reheated. If you don't have reheating facilities at work, leftovers stored at room temperature will be tasty by 1 p.m.

Another favorite of mine is a mixture of fresh veggies—carrots, celery, and cucumber, for instance—placed raw in the lunchbox with a small container of olive oil and vinegar or some homemade mayonnaise. This is very quick to prepare in the morning—you don't need to chop up the vegetables, just whack them in your lunchbox—and you can prepare the mayonnaise or dressing the night before.

Make it a team effort

Lastly, try to get your whole household involved. We all know what it's like to live in a busy home: kids, pets, neighbors, friends, roommates. Life can be chaotic. Trying to cook, garden, and keep chickens can create more madness if you aren't organized, or if one person alone is trying to make everything work. Include everyone in your mission: ask your kids to feed the hens (they'll love it), get your partner interested in the compost, make gardening a family activity. If you live with roommates, try to get them excited about an outing to a farmers' market or joining a co-op. Most people find that they get a lot of pleasure from doing these tasks, so it needn't be burdensome or arduous. Working together to get things started will make your frugavore experiment all the more productive and enjoyable.

STOCKING YOUR PANTRY

"It is thrifty to prepare today for the wants of tomorrow."

—Aesop

A TRADITIONAL KITCHEN PANTRY WAS STOCKED with all manner of preserves as well as ready supplies of herbs, vinegars, oils, legumes, and natural sweeteners. A well-planned pantry will save you so much time, worry, and money. You will be able to come home and prepare a healthy, inexpensive meal using just a few ingredients—no exhausting (and expensive) last-minute supermarket trips required. A basic lentil soup can be made with some dried lentils and herbs and some stock from the freezer. Chickpea soup can be thrown together with some dried chickpeas, some canned tomatoes, and a sprig of fresh herbs from the garden.

When stocking your pantry, don't feel you have to buy everything all at once. Start with the familiar, less expensive items first, but don't be afraid to experiment as you go. And when storing your purchases, remember that all pantry items are best kept in glass or ceramic containers and stored in a cool, dry place with minimal light.

Beans, legumes, & peas

Beans, legumes, and peas are extremely cheap and wonderfully nutritious. As a starting point try some dried chickpeas, lentils, red kidney beans, and split peas. With a stash of these in your pantry you will always be ready to make a wide range of soups, stews, and salads. See the "Legumes" section for recipes. Dried beans are the most cost-effective. They last for longer periods of time in the pantry and come with much less packaging. But if you are not the kind of cook who likes to plan ahead, it might be worth keeping a few cans of chickpeas and legumes on hand. They are a little bit more expensive, but they are quicker to prepare as they do not require pre-soaking or pre-cooking.

Fats & oils

Oil is essential for cooking as well as for dressing salads and other cold dishes. See the "Fats"section of this book for more information about modern industrial vegetable oils and why it's best to avoid them. Instead, in addition to animal fats (which are kept in the fridge), I recommend stocking your pantry with some good-quality cold-pressed olive and coconut oil.

Olive oil: This is the food of the gods. Enjoy it and purchase the best quality you can. Always look for extra-virgin, cold-pressed olive oil, and buy organic wherever possible. Olive oil should be stored in dark opaque glass bottles as it is sensitive to light.

Coconut oil: Coconut oil or coconut butter can be used for all types of cooking; it is especially well suited to high-temperature cooking such as frying and baking. Look for extra-virgin, cold-pressed, organic oil.

Flour

Whether you are an avid bread-maker or a very occasional baker, keeping a few different flours on hand will give you many more options when cooking. For truly fresh grains, you could consider buying your wheat berries whole and grinding them with a small kitchen flour mill. Alternatively, if you are purchasing ground flour, look for "stone-ground" and "organic" wherever possible. I try always to have a finely ground rye and spelt flour on hand. Because flour is sensitive to heat and light, it is best to keep it in the fridge or freezer. See the "Grains" chapter for more information.

Arrowroot powder, also known as tapioca flour: Made with the tapioca plant, this flour is an excellent thickener in soups and stews.

RICE

Rice is perfect for last-minute dinners, to flesh out leftovers, or complement the latest pickings from your garden. There are many sorts of rice available, but I generally prefer cooking with brown rice. This can be bought in bulk and stored in large containers. If you are worried about weevils, keep it in the freezer.

POLENTA

Polenta is essentially ground-up corn, and has been a peasant staple since the Middle Ages. Incredibly quick to cook, it's a wonderful last-minute dish and can accompany meat dishes, casseroles, and vegetables.

SAGO OR TAPIOCA

These tiny balls are extracted from the pith of sago palm stems. It's an excellent staple and has a very long shelf life. Sago pudding is a favorite dessert at our house.

COCONUT MILK & COCONUT CREAM

These are very useful ingredients for people who suffer from milk allergies. They can be added to custards or desserts in place of milk.

Seasonings

Salt: Every pantry should include sea salt, which is very different from normal table salt. It provides a wide range of minerals and stimulates digestion. It can be found at most health-food shops and supermarkets.

Dried herbs: Fresh herbs add the best flavor, but a supply of dried herbs is useful for when fresh herbs aren't available. Parsley, rosemary, thyme, oregano, and bay leaves are all indispensable.

Seaweed: Seaweed varieties including kombu, arame, and dulse flakes can be added to grain and seafood dishes and to bone broths and stocks. They are an excellent way to add exotic flavor and extra minerals. In some organic food stores they can be quite expensive, but you can find them at much better prices in small Asian supermarkets.

Spices: Dried cinnamon, nutmeg, ginger, allspice, and paprika are invaluable for both sweet and savory cooking.

Sweeteners: For day-to-day cooking and for when you crave a treat, you'll want a few sweeteners in your pantry. There are many healthier alternatives to refined sugar, including honey, molasses, rapadura sugar, and stevia. See the "Sweet Stuff" chapter for more details about their different uses. I like to keep a selection of these on hand, along with some brown sugar (which is cheaper than the healthier alternatives) for big cooking sessions.

Vinegars

Vinegar can be used to flavor salad dressings, is an essential ingredient in stock (it helps to release the calcium from the bones), and can be added to many soup recipes for a slightly tart flavor. It's also very handy for household cleaning. There are countless types of vinegar available, but the few listed here will get your pantry started.

Apple-cider vinegar: This is my favorite variety of vinegar, thanks to its tangy flavor and excellent nutritional value. Apple-cider vinegar can be used for all types of dishes, from salad dressings to soups and stews. Be sure to choose a brand that contains a "live mother culture" (the label may just say "with mother," which I always think is rather funny), and try to buy organic wherever possible.

Balsamic vinegar: This is a very popular vinegar, and useful for salad dressings.

Red or white-wine vinegar: These can be used for all the same purposes as apple-cider vinegar, but are not as nutritious. Wine vinegar is also good for poaching eggs and for adding flavor to salad dressings.

PRESERVES, PICKLES, & CONDIMENTS

A selection of jams, pickles, preserves, and condiments will enrich any pantry. The "Preserves" section of this book includes recipes for making your own. In addition to homemade jams and pickles, these store-bought preserves can come in handy.

Tomato passata or canned tomatoes: Preserved tomatoes can be added to soups, stews, and seafood dishes, and are especially handy when tomatoes are not in season. Ideally, look for good-quality glass bottles of organic passata or try the recipe for preserved tomatoes on page 317.

Fermented soy products (miso, natto, tempeh, and soy sauce): These traditional soy products can be found in Asian grocery shops, health-food shops, and supermarkets. Miso can be used to flavor soups and stocks, while soy sauce can be added to meat and vegetable dishes. Natto goes well in soups or with Asian noodle and vegetable dishes. When buying these products, look for "traditionally fermented" or "traditionally brewed" on the label. Also try to buy organic, as many soy products are now genetically engineered. Organic and traditionally fermented brands may be a little more expensive, but remember that a good bottle of soy sauce or packet of miso should last you several months, if not years.

Worcestershire sauce: This is useful for flavoring rich stews and bean dishes. Look for organic wherever possible, and check the label to be sure it contains no odd-sounding additives or preservatives.

Tomato ketchup: You can make your own ketchup (see the recipe on page 320), or buy it ready-made. Try to choose one that contains no preservatives or artificial ingredients. It goes beautifully with mince dishes such as hamburgers and meatloaf.

THE VEGGIE PATCH

"It is said that the effect of eating too much lettuce is 'soporific.'"

—Beatrix Potter

It wasn't all that long ago that almost every person who owned his or her own land also grew his or her own food. Every inch of backyard was used to its fullest capacity to grow anything and everything edible, or to support animals like poultry or pigs. You could walk past your neighbors' front yards and witness all manner of blossoming fruits, odd-looking speckled vegetables, and rabbit traps and pigeon huts ready to catch the evening meal. I sometimes dream of walking back in time into this world, where my own small patch of vegetables would be complemented by my neighbors'. All I'd have to do at dinner time is turn on my oven, get out my clippers, and walk out the back door.

Like many people today, I started my own vegetable garden as a way to deal with constraints of time and money. I was exhausted, shopping at organic food stores on the other side of town and coming home late to cook dinner. Our grocery bills were always high and I would arrive home late and frazzled. I might have a basket full of produce, but it would be too late or I would simply be too tired to cook.

Growing your own produce has a number of advantages. It will save you time, it is cost-effective, and (the best benefit of all) you will have access to the freshest, most nutritious produce.

Vegetable gardening does take time in the initial stages. You need to plan your plot, plant your produce, fertilize the soil, and keep an eye on its water and nutrient levels. But once your garden is established, it is actually very little work. And as I came to learn, the work that is required is generally enjoyable and often very sociable. At our place, it can easily become a morning-long affair with my mother, a cup of tea or two and some music blaring from the kitchen. Gardening isn't arugula science; it's a combination of knowledge and intuition, both equally important. Be prepared to give everything a go and you'll be a very good gardener.

If you are just starting out or have limited space, some herbs are the first thing you'll need. You don't need a garden bed: you can plant them in pots, in a window box, or even in a recycled container (we recently rescued an old sink from someone's dumpster and turned it into a herb garden—moveable and cost effective, and a conversation starter too!). A few fresh herbs growing on a windowsill or on the back doorstep are invaluable for home-cooking. Even the simplest of dishes can be improved with a few sprigs of fresh parsley or basil, and growing your own is much cheaper than buying them fresh every week. A bunch of fresh rosemary or thyme at our local super-market retails for three or four dollars, whereas a seedling from the nursery sells for a similar price and will last indefinitely.

If you have the space, a few pots of fresh greens—lettuce, arugula, silverbeet, or spinach—are also wonderful to have. These are super easy to grow, and with a handful of arugula and some pickled sardines or a fresh egg from your hens, you can have dinner organized in a flash. Greens really need to be fresh for optimum flavor; even if you have the time and money to shop for them regularly, they will taste much better freshly picked from your own backyard. You can buy greens as seedlings, if you want to keep things easy, or as seeds, which require a little more care. They will provide your garden with color and life and your kitchen with a convenient and highly nutritious food source.

Once you have a few basics under your belt, you can try venturing into a bigger garden plot, either in your own backyard or through a community garden. This only needs to be the size of a bathtub or two; in just a few square feet, you'll be able to grow a good proportion of your household's produce. An average suburban front lawn could supply a family of four throughout the year.

When it comes to choosing what to grow, there are countless options, but the easiest place to start is your local nursery. See what is most appealing and what is most suitable for your climate and season. Try to keep your garden as varied as possible and rotate your crops each season. Nature is built around the principles of biodiversity, so by regularly changing your crops and planting different species of plants together, you'll add nutrients to the soil and decrease the risk of pest attacks. Including some animal life in your backyard—be it by keeping chickens, or adding manure to your compost heap—is another good way to replenish the soil.

If you build a healthy soil, water regularly (and do some clever drought-proofing), and plant seasonally, before you know it your garden will be thriving.

PLANTING IN POTS

If you only have a small garden, courtyard, or balcony, planting in pots may be your best bet. This could set you back a couple of hundred bucks at many nurseries, with their wide array of pots and potting mixes. There's no need to spend a fortune on designer containers and fancy soils, however; here are some frugal tips for getting things started more cheaply.

Choosing your pots: Wherever possible, choose pots as big as you can fit in your space. The more room your plants have, the further their roots can grow. Heavy pots (such as ceramic, terracotta, or metal) and plentiful soil will protect your plants against extreme temperatures. Ceramic or terracotta pots can be expensive, however. To save money, look for pots at thrift shops and recycled hardware stores, and experiment with what you already have or can scavenge. My most treasured pot is a bright orange barbeque base that I removed from a dumpster. My boyfriend delicately drilled two holes in the bottom, and it now holds a few varieties of lettuce and several clumps of parsley. We've also collected large stainless steel containers, old pig troughs, and a baby's bath, each of which we've filled with homemade potting mixture and turned into a thriving vegetable bed. If you can get your hands on them, halved wine barrels make excellent pots—they are big and deep, and the timber provides insulation (and, dare I say it, a nice winey aroma!). Wine barrels are often sold at nurseries or hardware stores, and can be considerably cheaper than ceramic pots. Even better, vineyards often sell used ones for next to nothing, or may even be willing to give them away for free.

Plastic pots are another option for the low-budget gardener. They don't insulate the plant as well as ceramic or steel, and I tend to be anti-plastic as a general rule—but a plastic pot is better than no pot and no garden. Some nurseries give away used plastic pots for free.

Make sure you keep your pots in a sunny spot. As a general rule, plants need at least six hours of sunshine per day, particularly during winter. Also try to elevate them off the ground (for instance with feet or a tray) so that the water can easily drain from the bottom. You'll need to water your pots regularly and keep them protected during hot, dry summers.

Potting mixture: Vegetables don't tend to do so well in pots with just standard garden soil. You'll need to make up a potting mixture to help your plants thrive and provide adequate drainage. Some gardeners prefer to make their own potting mixture, or to use a combination of bought mixture and a few boosters from their own backyard. Here's a rough recipe I use regularly. The basic ingredients can be bought at most nurseries or hardware stores:

  2 parts commercial potting mixture or
     healthy topsoil from your garden
  1 part compost or worm castings, mixed with
     bark or garden clippings in a 50:50 blend
  1 part coarse river sand

If you're trying to make your commercially bought mixture or topsoil go further, you can reduce the amount so that your recipe is 1 part potting mixture, 1 part compost, and 1 part sand.

Potting fruit trees: Fruit trees can grow well in pots. Look for dwarf varieties, as they provide an abundance of fruit and don't grow too big. You should easily be able to fit a dwarf citrus tree in a large pot or wine barrel. Potting

mixture for citrus trees is different from that for vegetables. They require additional sand to provide extra drainage and encourage root growth. Here's my recipe:

  1 part garden topsoil
  1 part compost or worm castings, mixed
     with bark or garden clippings in a 50:50
     blend
  2 parts coarse river sand

Nature's recycling systems

My own garden only started to thrive when I began to recycle our kitchen waste back into the soil. When my brother moved, he gave me a worm farm he didn't want anymore. I dutifully put all our kitchen waste in it, and it soon started to produce an amber-colored worm juice. My brother had advised me to put this on the garden, so I diluted it in a watering can and poured it onto a couple of silverbeets I was growing (unenthusiastically, I'll admit: I was still a bad gardener at this stage!). Within a week, the plants looked like they had been given a dose of anabolic steroids. My confidence grew quickly: I started planting more seeds and feeding the worm farm more regularly. I used the solid worm castings as potting mixture for new plants and spread them on the garden. I knew very little about proper soil ecology or plants generally, so I was relying on intuition. The worm farm seemed to make sense and the nutrients allowed the plants to really thrive. Within six months, with a growing sense of confidence, I had ripped out our front lawn and was growing zucchini, pumpkins, and tall stalks of artichokes and leeks. I am not a clever gardener and didn't expect things to go this well; it really all started with the worm farm and a good healthy soil. The rest was remarkably easy.

Nature has a clever way of recycling itself; the process of plant death, breakdown, and regrowth is an intrinsic part of our natural environment. Plants have a well-defined life cycle: they grow, blossom, ripen, then break down and fertilize the ground around them. Composting and worm farming are not only great ways to dispose of waste; they also let you harness these natural processes to improve your soil. If vegetable and fruit scraps end up at the land fill, this plant cycle creates one of the most environmentally damaging gases—methane—which is produced as plant matter breaks down and is released directly back into the atmosphere. So by composting, you'll be helping the environment as well as your own little patch of earth. In fact, it has been estimated that if every person had their own worm farm or compost heap, it would reduce our waste output by 1 ton per person per year. Imagine what we could achieve if more communities, corporations, and local councils embraced this—we could see local drop-off points where people discarded their kitchen waste, and it could be used to fertilize parks, local gardens, and community vegetable patches.

Composting and worm farms allowed me to create a mini-ecosystem in our home and garden. Our fruit and vegetable scraps go to our worm farm, along with layers of shredded newspaper and tea leaves. Onions and citrus fruit (which the worms don't like) go to a bokashi bucket, which uses micro-organizms to ferment them; I can later plant them in the garden or at the bottom of pots. Meat scraps and bones are given to our dog or buried in the garden (deep enough that the dog can't dig them up!), where they fertilize the soil. The chickens also love leftovers—porridge, stale bread soaked in water, and greens—so there are always treats for them. I often see people struggling to nurture plants that won't thrive or spending loads of money on mulch and fertilizers. Gardening doesn't need to be so complicated! Start simple: recycle your waste, get a good composter or worm farm, and the rest should go from there. Below, I've outlined a few different ways to get started with composting.

COMPOST HEAPS

Compost heaps are the most time-tested and traditional way to recycle your waste and fertilize your garden. They take up more space than a worm farm and are generally best suited for medium to large backyards. Don't be dissuaded by this, however, as it's well worth the space it requires; compost can be used as a mulch, potting mixture, or liquid fertilizer, or to prepare a garden bed for a fresh season of planting.

The two most popular designs for compost bins are the traditional plastic bin, which sits upright in a corner of your garden, and the tumbler bin, which looks like a plastic wine barrel supported by two timber posts and is easier to rotate and swing. Tumbler bins are rat-proof and easier to move. The only drawback is that they are usually a lot more expensive, often a couple hundred dollars. If you are handy with a tool kit, you could construct your own (have a look around online for instructions).

More traditional compost bins, which sit upright and stationary on the ground, are much cheaper (usually about fifty dollars from gardening or hardware stores). I've also seen some good buys on eBay and through local councils.

Handy home-gardeners, however, can easily create their own compost heap with a few basic components. With some timber logs, a two-yard square area can be sectioned off and used for your compost heap. Any large, weatherproof container will also work well. My mother uses a large recycled corrugated-iron water tank; she can easily get inside and stir the compost with a pitchfork. Another friend of mine has converted a large bathtub into a compost heap; it's affectionately known as "the coffin."

When it comes to choosing a spot for your bin, remember that compost needs plenty of warmth and sunlight for fermentation to take place. It's also a good idea to situate your bin as close to the kitchen as possible, so that you actually use it!

Developing your compost: For a healthy compost heap, you need to balance wet and dry ingredients to ensure the right level of moisture. This is easier than it sounds—really it boils down to common sense. Dry ingredients include any type of dried mulch (garden leaves, shredded paper, hay, or straw) and garden cuttings. Wet ingredients include freshly dug weeds, manures, fruit and veggie scraps, and lawn clippings. A healthy balance of these will provide the right environment for micro-organisms to thrive and prevent the heap from becoming wet and stinky or dry and dusty.

If you are using a tumbler bin, you can simply monitor the ingredients you put in; they will all be mixed together when you spin the plastic tub. If your heap is based on the ground, try to layer the different ingredients like lasagna, so that they are all spread evenly throughout the heap. You can have a layer of vegetable scraps, followed by sheets of newspaper, followed by manure, then some more paper, and then some straw or mulch. For some extra nutrients, add a bit of basalt rock or a few spadefuls of garden soil—this will reinvigorate your compost with a good dose of minerals and healthy micro-organisms.

For it to break down into compost, the waste needs to be tossed or rotated regularly. Tumbler bins can simply be spun whenever you happen to walk past. If you make your own bin, just use a pitchfork to toss the compost at least once a month.

Your compost is ready for your garden when it is a rich, dark color, crumbly in texture, with a warm, earthy smell. If your compost is on the ground, this can take up to eighteen months. If it is in a tumbler, it should only take three to four weeks. Some keen gardeners like to have two separate bins on the go; while one is fermenting and breaking down, the other can be filled with new ingredients.

Compost ingredients: All of the following things can be added to your compost heap.

If you cut up your ingredients as small as possible, they'll break down more quickly:

- Uncooked fruit and veggie scraps
- Coffee grounds and tea leaves
- Shredded newspaper
- Old garden mulch, straw, lawn clippings, autumn leaves, and prunings
- Fireplace ash
- Cow, sheep, horse, or chicken manure
- Dolomite lime and basalt rock
- Garden soil

Ingredients to avoid:

- Cat or dog droppings
- Cooked food (I do occasionally add small amounts of cooked vegetable matter, as it will break down fairly easily if the compost is strong and healthy. As a general rule, however, cooked scraps are best avoided)
- Meat matter, whether cooked or raw, which can breed bad bacteria and parasites

Using your compost:

- As mulch: spread your compost around your garden plants, applying it up to 4 inches deep.
- As liquid fertilizer: in a bucket or tub, add one part compost to three parts water. Give it a good stir, then leave it for two to four days to ferment. Apply this liquid to the plants as a tonic.
- As potting mixture: compost can be combined with other ingredients to produce a rich and healthy potting mixture.
- To prepare a garden bed: reinvigorate your soil with plenty of compost just before a new season of planting.

Worm farms

I love worm farms. They are usually small—less than 3 feet in diameter—which means that they can fit on even the smallest apartment balcony. They can be fed uncooked kitchen waste (vegetables and fruit peelings) as well as some garden waste and small quantities of newspaper scraps.

Worms are integral to a healthy soil environment; they recycle and transform organic waste matter into useable, nutrient-dense soil humus. You want to encourage as much worm activity in your garden as possible, and having a worm farm is usually the easiest way to achieve this.

You can buy a worm-farm kit from most hardware stores and nurseries. They usually retail for between eighty and one hundred dollars, but you may find them cheaper on eBay or at recycled hardware outlets. Some local councils sell them at a reduced price as an incentive for people to recycle their waste.

A worm farm consists of several layers of plastic trays. The worms migrate between the different levels as they work their way through your scraps. Most ready-made worm farms contain three moveable shelves. Farms can be circular or rectangular shaped, and they are usually on stilts so that they are out of reach of mice and rats. As the worms eat their way through the scraps in the bottom tray, you start adding your scraps to the next and they will gradually migrate upwards to the new food source. The worms in worm farms are different from earth worms—they are usually Indian blues, tiger worms, and red wrigglers. They can be bought from nurseries and hardware stores. Alternatively, if you have a friend with an already thriving farm, ask if you can take a container of worms from their farm to start your own.

If you are strapped for cash, a homemade worm farm is cheap and easy to make. All you need are two large polystyrene boxes (about 20 by 20 inches) and a well-fitting lid for one of the boxes. Make sure that the boxes can fit neatly one on top of the other. In one corner of the bottom box, drill a hole so that the liquid can drain out. Poke at least twenty holes in the bottom of the top box, so that the worms can travel between boxes without difficulty. On the bottom of the farm, place a layer of newspaper or hessian material. Start your worms on the bottom layer, with a little bit of food, then build up their intake as their population grows. When they are eating well on the bottom layer, transfer half of the material from the bottom layer to the top layer and continue feeding on both levels.

Worm farms produce a special juice (commonly referred to as "liquid gold" by home gardeners), which comes out of a tap on the bottom layer of the farm. You can dilute this juice to use as fertilizer on your garden. You can also use the worms' solid waste (or "castings") as potting mixture. Anecdotally, biodynamic farmers have told me that worms can be used to correct imbalances in a plant's nutrition. For instance, if you are growing cabbage and it's not looking too healthy, place a bunch of the ailing plant in your worm farm. The worms will break it down and create the nutrients the plant is lacking; add their "worm juice" to your sickly cabbage plant and it should improve.

Getting started: When starting a new worm farm, it's a good idea to provide some healthy bedding at the bottom, so that your worms can settle in to their new environment. I recommend placing a few layers of newspaper or a hessian sack on the bottom tray of the farm. Then add the worms, some feed, and a spadeful of your garden soil.

Temperature and positioning: Worms like to be kept in a cool, dark place. If they get too hot (above 98.6°F) they can die. You will also find that their activity will markedly increase during wintertime, but slow down during summer. During the peak of summer, keep your worm farm in a shady spot and regularly tip cold water in to prevent them from drying out. A cold wet hessian sack on top of the farm will also help to keep them cool.

Feeding: Worms like to be fed regularly (at least once a month) and they like to be able to get through their feed at their own pace. If you feed them too much too quickly, they may not be able to keep up and some of your waste may go bad, disrupting the ecology of the farm. Worms love to eat uncooked kitchen waste (fruit and vegetable scraps), tea leaves, coffee grounds, newspaper clippings, and any other organic matter you can drum up. If you can, cut up your scraps into small pieces before adding it to the farm. It will be eaten and broken down much more quickly, preventing it from going bad.

Like compost heaps, worms prefer to get through their food in "layers"; this protects them against extreme temperatures and balances the wet and dry ingredients within the farm. I usually alternate layers of kitchen waste with thin layers of newspaper (one or two sheets should do it). Every now and then I also throw in some basalt rock and perhaps even a little garden soil. Using my intuition, I add food depending on how well they are going. If their activity is high and the worms seem to be thriving, I'll throw in as much food as I can. But if they are struggling, reduced in numbers, or their activity is slow (as in the peak of summer), I'll add only a small amount of food, plus thin layers of newspaper, some soil, and some basalt rock.

Worms cannot eat any meat or oily foods. They also don't like dairy products or acidic foods such as onions, leeks, and citrus fruits.

Troubleshooting: The two main problems people encounter with worm farms are the worms dying out or the worm farm "rotting" and stinking. The first problem occurs if the temperature is too hot for the worms to survive, or if they are not fed appropriate food. The second problem occurs if you feed the worms too much, too quickly. To prevent this, avoid inappropriate foods and measure your worms' food intake, feeding them greater quantities as their population grows. If you encounter fruit flies or bad smells, it might be that the farm has become too acidic. This can be corrected by adding some potash, lime, or dolomite. You will also need to add a few more dry ingredients, such as newspaper or garden soil, to balance things out.

BOKASHI BUCKETS

Bokashi buckets are small enough to sit on your kitchen bench, usually about 1 foot by 1 foot. They can break down anything—food scraps, meat, dairy products, and vegetable matter. They come with a special type of sawdust that contains active micro-organisms, which break down your waste. You can plant the broken-down scraps in your garden or pot plants to nourish the soil. Bokashis only break down scraps; they don't create the same rich fertilization you get from a compost bin or worm farm. They are, however, a very compact solution, and can take just about anything. I would suggest using a bokashi bucket if you are short on space or time, or as a way to dispose of meat and citrus scraps if you already have a worm farm or compost heap.

Bokashis are at the more expensive end of home recycling, usually selling for eighty or ninety dollars. The bokashi sawdust will set you back about ten dollars per refill. Some of my more innovative home-gardening friends have made their own bokashi buckets by drilling holes in large plastic containers, securing a lid, and buying just the special sawdust—so don't be afraid to experiment.

GARDEN BURIALS

If you don't have the space to compost or don't have the money to invest in any of the aforementioned composting systems, there is another alternative. Collect all your kitchen scraps, dig a hole, and bury them in your garden. If the hole is deep enough they will not be dug up by animals and they will nourish your soil in the long term. Uncooked kitchen scraps work best for this method. If you are using cooked scraps or meat, you may want to throw in some sawdust or bokashi flakes to ensure they break down properly and don't get stinky. Similarly, if you have pot plants, I would suggest planting your food scraps at the bottom of your pots, adding some sawdust or bokashi flakes, then covering them with a thick layer of soil before you plant your seeds or seedlings.

GROWING FROM SEED

I'll never forget the first time I discovered a large cantaloupe sprouting uninvited from our veggie patch, or a pumpkin vine unexpectedly edging its way across the lawn. Self-sown seeds spring frequently from the nurturing environments of worm farms and compost heaps, but the joy of witnessing those first surprise buds never ceases to amaze me.

Almost all fruits and vegetables can be grown straight from the seed, and they'll add another dimension of creativity to your backyard veggie patch. In true frugavore style you can throw a rotten tomato into a greenhouse basket and watch it spring tall shoots, or allow the leftover seeds from pumpkin soup to germinate in the warm summer soil and creep up a fence, creating ripe, orange fruit. Or simply plant the seeds in your garden during a warm period (preferably spring) with some worm solids or compost around each seed and let nature take its course. If you buy heritage varieties of seeds you can grow interesting and odd-looking plants such as black tomatoes, purple carrots, and five-color silverbeet. You will also, in your own clever way, be promoting our planet's biodiversity, as well as saving money and reducing waste.

You can grow plants from seed either by sowing them directly into the ground or by planting them in planters in a greenhouse. Direct sowing works best for plants that don't mind cooler temperatures, such as root vegetables. Planters are required for seeds that need warmer temperatures to germinate. Because they are small and shallow, planters tend to warm up faster than the ground, while the greenhouse provides a protective environment free of wind, birds, and dryness.

A mini-greenhouse is a good guarantee that your seeds will grow: it provides the most nurturing environment possible. At large hardware stores you can buy yard-wide greenhouses with three or five shelves for less than fifty dollars. If you don't have the money or space for one of these, you can plant your seeds in recycled polystyrene boxes, which will insulate them from the heat. However if you grow your veggies in planters, try to keep them in a warm, protected area.

Some easy seeds to sow

PUMPKIN: You can use the seeds from any store-bought organic pumpkin. They germinate easily, except during heavy frosts.

TOMATOES: Seeds from any organic variety of tomato will work. Sow your seeds in containers six to eight weeks before you intend to plant them in your garden. In warm, frost-free areas where soil temperatures are above 59°F seeds can be sown directly into the ground.

MELONS: You can use seeds from home-grown or organic store-bought melons. Melon seeds require lots of heat to germinate. They sprout easily in containers or in warm, well-composted soil.

GARLIC: Replant unused garlic cloves, pointy-end up, about 2 ¾ inches deep in well-composted soil. They do not require containers and can be sown in autumn.

ONIONS: Onion seeds can be sown directly into the garden, but it is generally easier to sow them in containers. Most onion varieties can be sown at any time of the year.

POTATOES: Old, saggy potatoes can be transformed into new potatoes by planting them in your garden bed. Each planted potato should yield ten new potatoes. Plant them in spring, about 4 inches deep, with plenty of fresh manure. They will take about 120 days to produce a new crop. All parts of the potato plant are poisonous except for the tubers (the part we eat), which should have no green tinge.

WHAT TO PLANT WHEN

Choose vegetables and plants according to your local climate and soil conditions. Here is a general guide to what to plant when:

SUMMER (June through August): tomato, eggplant, peppers, melons, cucumber.

AUTUMN (September through November): silverbeet, greens, beetroot, garlic, celery, fava beans.

WINTER (December through February): potatoes, broccoli, cabbage, fava beans, leek.

SPRING (March through May): string beans, peas, zucchini, lettuce, spring onions, pumpkin.

BUILDING & MAINTAINING YOUR SOIL

If you want to give your plants the best start, you need to begin with healthy soil ecology. This can be done by mulching, fertilizing, and removing surrounding weeds, as well as consistent upkeep and watering. Don't feel you need to go out and buy a lot of products to achieve this; it's actually a lot simpler than you might think. If you have a compost heap, worm farm, or bokashi bucket, you're off to an excellent start—many gardens get by beautifully with one of these and nothing else. Listed here are some other ways to build and maintain healthy soil.

Each garden is different, and different fertilizers will suit different soils. An easy way to test whether a particular fertilizer is suitable for your soil is to take a teacup of worms and soil from your garden bed and mix in the new fertilizer. If the worms continue to wriggle around at the bottom of the teacup, they probably like the fertilizer. If they don't like it they'll go crazy, trying to get out of the teacup as quickly as possible. This test is especially handy if you're buying commercially sold fertilizer, which may contain traces of chemicals.

Basalt rock: This is a rock mineral sold at most nurseries. It contains important trace minerals necessary for plant growth. It can also be added to compost heaps or worm farms to facilitate growth and fermentation and balance out pH levels.

Blood and bone: This is full of minerals, including slow-release nitrogen, calcium, and phosphorus. A frugal alternative to buying commercial blood and bone is to ask your local butcher for some bone shavings from the back of his workroom floor. This will do the trick just as well.

Dolomite lime: Dolomite lime is a naturally occurring rock mineral that boosts plant growth and alkalizes the soil. Dolomite can be added in moderate amounts throughout the plant's life cycle. Read up on the plants you're growing first, however, as some plants—namely those that prefer an acidic soil, including azaleas and blueberries—don't like it.

Manure: Sheep, chicken, and livestock manure are all useful fertilizers, but they need to be added to your compost heap first so that they can break down and ferment before being added to your soil. You can buy them from nurseries or from the side of the road in rural areas or near stables, usually for two or three dollars per bag. Some plants really thrive on animal products. Rhubard is a classic example—a little manure and you'll have stems as thick and strong as tree stumps. The best time to add animal products to your soil is at the beginning of spring or autumn when you are about to plant new seedlings.

Mulch: This can include pea straw, straw, lucerne hay, autumn leaves, or even newspaper clippings. Mulch provides a protective covering for your plants against extreme temperatures and other climatic conditons. It also adds organic matter to the soil and encourages worms. Pea straw and hay are probably the best choices for their rich organic matter. Autumn leaves collected from your lawn can also be added straight onto your garden bed, or mixed into the compost bin. If you are worried about autumn leaves or other forms of mulch flying away, you can weigh them down with a couple of sheets of newspaper and a brick. Otherwise, just alternate sheets of newspaper with soil from your garden. It won't have the same nutritional properties as straw or garden leaves, but it will provide protection for the worms and encourage them to grow and breed. Mulch is best added during planting time, or at the beginning of extreme temperature periods such as summer or winter.

Mushroom compost: This is the residual waste generated by mushroom farmers. It's a great source of nutrients and is also a good soil conditioner. Only add mushroom compost in moderation as it can raise the soil's pH levels and create too alkaline an environment.

Seaweed: This provides the soil with important minerals, including iodine. When I was growing up, we used to rake this from the side of the beach, but I'm not sure if this is still legal. You can buy seaweed from most nurseries.

Lead testing: One thing I didn't take into consideration when I started my own garden, but which I now understand to be important, is lead testing. If you are establishing a garden for the first time and live in an inner-city area, it can be worthwhile testing your soil for lead levels. If your soil has high levels of lead, you will need to build your garden elevated from the soil or choose large pots to avoid contamination of your produce. Lead-testing kits can be bought from most major hardware stores.

Long-term maintenance

Once you've got your vegetable garden established, there is very little you need to do, other than providing your plants with ongoing care and maintenance.

Regular weeding: This shouldn't be a difficult job. Clear out all the weeds surrounding your plants, but be careful not to put them back in your compost or worm heap, as they may sprout and re-grow. We usually leave ours on the footpath to shrivel up in the sun, and only then put them back on the garden bed. Weeding is most important during wintertime, when plants are competing for sunlight.

Regular fertilization: The best time to add nutrients to your soil is just before you plant—i.e., the beginning of autumn or spring, when you are about to plant your new harvest for the upcoming season. After this, a little bit here and there whenever your plants look like they need a bit of love is always a good thing.

Keeping bug-free: There are many different ways to get rid of bugs from your veggie patch. Here are a few I've learned along the way.

Bugs can flourish when the soil becomes too acidic. Throw on some dolomite lime and some compost or fertilizer to boost your plants' immunity.

Let your chickens loose in your veggie patch for a day or two. They'll get rid of any bugs, but you may lose a few vegetables as casualties. Consider it a two-day blitz, after which you can re-plant and recover.

To exterminate slugs, place a ring of salt around the plants they seem to favor. You can also fill a small bowl with beer. The slugs will crawl into the bowl, become intoxicated, and die.

Placing netting over plants can often solve

the problem of larger bugs (and also ward off birds).

Protection against heat and cold: Some plants can't cope in extreme climates—be it the heat of summer or the frost of winter. You can provide protection by draping them with a hessian cloth (which can block out sunlight, or insulate against the cold) or bringing them indoors.

Watering and drought-proofing: As our climate is becoming hotter and drier, we need to think of innovative ways to keep plants hydrated and cool. Plants love to be watered regularly. In the heat of summer, vegetables really need a drink at least every couple of days.

Pea straw, straw, or mulch are an excellent investment as they allow the plants to retain moisture around the roots, and hence not require too much watering.

You can also save your shower and bath water using a bucket and tip this over the vegetables after you've washed. Water tanks and recycling systems are also excellent investments.

Lastly, get a citrus tree! If you are a frugavore with a backyard, a citrus tree is an essential water-saving measure. Not only do they look lovely and require only minimal watering, they also provide a staple ingredient for home-cooking. Imagine all those ripe blood-oranges you could grow, or those lustrous lemon trees dripping with fruit. Citrus trees thrive on nitrogen, and the best source of nitrogen is … wait for it … urine. If you can convince the male members of your household to pee on your tree, you will save a lot of water and get beautiful lemons for your cooking. Just think, if every flush of the toilet uses between ¾ and 1 ½ gallens of water, imagine how much we could save just by peeing on our citrus trees every day. In no time, I'm sure, we'd be the marmalade and lemon-tart capital of the world …

Heirloom Roast Vegetables

I love the heirloom varieties of carrots—red, purple, white, and orange. They take a little longer to cook but are worth the extra wait. You can buy heirloom varieties at some organic shops, but the easiest way to enjoy them is to purchase the seeds and grow them yourself. Of course, plain old orange carrots will also do just fine.

Preheat your oven to 350°F. Drizzle some cooking fat into a baking tray.

Scrub the vegetables and remove the stalks from the carrots. Cut the onions into wedges. Boil the carrots in a little water for about 10 minutes or until they are slightly soft.

Drain the carrots and arrange them in the baking tray. Add the onions and toss through the fresh herbs and seasonings. Bake for 30 to 45 minutes, or until cooked through and crispy.

Preparation time: 10 minutes
Cooking time: 45 minutes
Serves: 4

Ingredients:
fat for frying
6 to 8 medium carrots
3 onions
¾ tablespoon fresh rosemary
¾ tablespoon fresh thyme
salt and pepper

HUMBLE BAKED POTATO

Potatoes are one of the most cost-effective vegetables to grow at home. Even when I was renting a house with no garden, I found a large plastic pot, filled it a quarter full with soil, added some old organic potatoes that were beginning to sprout, and watched them grow.

Baked potatoes can be a hearty meal in themselves. Try them with melted cheese, bean salad, sauerkraut, or yogurt and dill. The possibilities are endless, but this is my personal favorite.

Preheat the oven to 390°F.

Wash the potatoes and poke a skewer through their centers a few times. These airways will help the potatoes to cook evenly. Place them in the oven and bake for 50 minutes or so (depending on the size of the potato).

While the potatoes are baking, caramelize your onion. Put 1 teaspoon of fat in a frying pan and place it over low heat. Thinly slice the onions, then add them to the pan and simmer, stirring occasionally, for 25 minutes. They should become sweet and caramel-colored, but shouldn't burn. When they are ready, remove them from the pan and set them aside.

Preparation time: 5 minutes
Cooking time: 50 minutes
Serves: 2

Ingredients:
2 large potatoes
2 or 3 medium onions
fat for frying
2 slices bacon, cut into bite-sized cubes
1 generous handful flat-leaf parsley
¾ tablespoon freshly chopped dill
½ cup plain yogurt
1 tablespoon grated parmesan cheese
fresh butter to serve

While the pan is still hot, add the bacon and fry it until it lightly browns.

Next, finely chop the parsley and place it in a small bowl.

Finely chop the dill and combine it with the yogurt.

When your spuds are ready, you can layer the toppings however you like. I usually like to add a dollop of butter first, followed by the onions and bacon, then a dollop of yogurt, a spoonful of the parsley, and a sprinkling of parmesan cheese. Do as you see fit!

BUBBLE & SQUEAK

Bubble and squeak is a traditional English dish made with shallow-fried leftover vegetables. It became popular during World War II as it was an easy way to reuse leftovers during food rationing. In has also been referred to as "bubble and scrape," as it can be made using whatever leftovers you can scrape together. Don't be put off by its wartime origins, though: brussels sprouts, potatoes, and duck fat are a match made in heaven, and you don't have to be on food rations to enjoy it.

Clean the potatoes and chop them into small rectangles (about ¼ inch by ¾ inch).

Heat a large, heavy saucepan over medium heat. Fry the onion, garlic, sage, and duck fat for a few minutes, then add the potatoes. You may need to reserve some of the duck fat and add it to the pan gradually, to keep the potatoes from sticking. Cook them for about 10 minutes, until they are crispy on the outside but still firm in the middle.

While the potatoes are cooking, remove the outer skins from the brussels sprouts and cut the sprouts into quarters.

Preparation time: 10 minutes
Cooking time: 15 minutes
Serves: 6

Ingredients:
1⅓ pounds waxy potatoes
1 medium onion, finely chopped
1 clove of garlic, crushed and sliced
1 small handful sage, finely chopped
1 spoonful of duck or goose fat
2¼ pounds brussels sprouts
1½ cups water
salt and pepper
olive oil

When the potatoes are ready, add the sprouts to the pan and stir them through. Ideally, the sprouts should go to the bottom of the pan, where they'll soak up the most liquid, but don't panic if they go everywhere.

Turn up the heat, pour in the water, and secure the lid of the frying pan. Perform the "fry-pan shuffle," shaking the pan so that the sprouts move around a bit and cook evenly. They should take 4 or 5 minutes to cook in the steam from the water. Take care to stir the mixture once or twice to make sure it's cooking evenly and not sticking to the bottom. If there is any excess liquid once the brussels sprouts are cooked, simply remove the lid and let the excess water evaporate.

Season with salt, pepper, and a dash of olive oil.

WILD GREENS PIE

I became obsessed with spinach pie after traveling through Greece, where horta—their trademark bitter greens—can be found at most restaurants and delis. I am sure this recipe is nothing like the original, but I've tried to capture the things I love about traditional Greek pie as nearly as possible (any Greeks reading this, you can stop laughing now). You can use any garden greens for this pie: silverbeet, spinach, kale, cavolo nero, or the traditional Greek horta. Leafy greens are easy to grow at home, and should be top of the list for any first-time gardener. They are just about foolproof and you should only need to plant them once a season—if you pick at them regularly, they will last you a good part of the year.

Preheat the oven to 350°F.

Prepare your pastry. Use ⅔ of it to line a 9-inch baking tin and bake "blind" for 15 minutes. Remove the "bake blind" filling and return the pastry to the oven for a further 5 minutes, allowing it to lightly crisp. Keep the rest of the pastry aside for later; it will become the lid.

Chop the greens into 1-inch pieces. Dice the zucchini and cut the broccoli into small florets. Sauté these vegetables in

Preparation time: 15 minutes
Cooking time: 45 minutes
Serves: 8

Ingredients:
2 quantities oatmeal pastry
   (see the recipe on page 252)
3 large bunches of greens
2 large zucchini
1 head of broccoli
8 ounces feta cheese
3 eggs
1 teaspoon freshly grated
   nutmeg
juice of 1 lemon
¾ tablespoon olive oil
salt and pepper

batches in a large frying pan with a dash of water, then drain them of any excess fluid.

In a large mixing bowl, combine the feta cheese, two of the eggs, nutmeg, lemon juice, and salt and pepper to taste. Add the cooked vegetables and stir well.

Transfer the filling to your pastry shell and spread it out evenly.

Roll out the remaining pastry to make a lid and place it over the pie. Whisk the remaining egg with a fork and brush it over the lid.

Bake the pie for 35 to 40 minutes, or until the lid is golden brown.

broccoli

tomato & onion pie

This is a terrific last-minute dish, which I've borrowed from my mom. The idea is to use whatever you have on hand, so don't worry if the bread is a little stale or if you have some other variety of cheese in the fridge. Provided you use the ripest, most succulent tomatoes you can find, your pie will be delicious.

Preheat the oven to 350°F.

Thinly slice the onions and place them in the bottom of a deep baking dish. You don't need to add any fat; just pop them in the oven and bake for 10 to 15 minutes while you prepare the remaining ingredients.

While the onions cook, thinly slice the tomatoes. Rip the bread into tiny breadcrumbs, or whir it in the food processor for a minute or two. In a small bowl, combine the breadcrumbs with the fresh herbs, cheese, olive oil, and salt and pepper to taste.

Remove the baking dish from the oven and layer the tomatoes over the onions. Sprinkle the bread and cheese mixture over the top, then return the dish to the oven and cook for a further 30 minutes, or until the tomatoes are soft and the cheese is lightly browned and crispy.

Preparation time: 10 minutes
Cooking time: 25 minutes
Serves: 6

Ingredients:
4 or 5 medium onions
4 or 5 large, ripe tomatoes
2 slices sourdough bread
1 teaspoon finely chopped
    fresh thyme or rosemary
1 tablespoon finely chopped
    parsley
¼ cup grated parmesan cheese
¼ cup grated tasty cheese
¾ tablespoon olive oil
salt and pepper

cauliflowers with bacon

I used to think of cauliflower as a rather dull vegetable, but exploring different varieties made me realize how wrong I was. Heirloom cauliflowers are a real treat—on a recent trip to a local farmers' market I came home with a purple, white and yellow beauty. Cauliflower goes deliciously with bacon and this is a very easy and impressive side dish, especially if you can find a cauliflower in an unusual color. If you don't have access to heirloom varieties, use the standard white one; it's still delicious when cooked this way.

Cut the cauliflower into quarters and boil or steam it for 5 minutes, or until lightly cooked and "al dente." Remove it from the heat and drain it of any liquid. When it has cooled a little, cut the cauliflower into bite-sized florets.

Heat a frying pan over medium-to-high heat and add the bacon. Allow it to brown, turning it once or twice. When the bacon has browned, add the cauliflower and stir well so that it is coated in the juices from the meat.

Transfer the cauliflower and bacon to a serving bowl. Add the parsley, lemon juice, salt to taste, and a little olive oil. Toss well and serve immediately.

Preparation time: 5 minutes
Cooking time: 10 minutes
Serves: 6 (as a side dish)

Ingredients:
1 large cauliflower
2 slices of bacon, finely chopped
1 small handful flat-leaf parsley, finely chopped
juice of ½ lemon
sea salt
olive oil

Garden salad

With a few pots of this and that or a small garden bed, a fresh salad is never far away. When you're growing your own produce, salads are a magnificent way to celebrate every twist, turn, and oddity in the plants from your garden.

Preheat your oven to 350°F. Lightly grease a baking tray with a little cooking fat.

Peel the beetroot and cut it into bite-sized cubes. Place them on the baking tray and sprinkle with sea salt. Bake in the preheated oven for 25 to 30 minutes, or until golden and crispy.

Meanwhile, shred the lettuce into a large salad bowl. Finely slice the fennel, cut the artichokes into small slivers, and roughly grate the carrot. Add these to the lettuce, then toss through the vinegar and olive oil.

Remove the beetroot from the oven and allow it to cool to room temperature. When it has cooled, arrange it on top of the salad and serve.

Preparation time: 5 minutes
Cooking time: 30 minutes
Serves: 6

Ingredients:
fat for frying
1 plump beetroot
sea salt
2 large heads of lettuce
2 large fennel
2 preserved artichoke hearts
1 large carrot
1 teaspoon white-wine or
    apple-cider vinegar
1½ tablespoons olive oil

THE CHICKEN & THE EGG

"Regard it as just as desirable to build a chicken house as to build a cathedral."

—Frank Lloyd Wright

I WOULD LOVE TO SEE A DAY WHEN EVERY household with its own backyard has a few chickens running loose, every apartment block has its own henhouse, and every farm has fresh eggs available for its neighbors. Chickens can play an important role in your garden's ecosystem: plants thrive when fertilized by hen droppings. Hens also provide excellent eggs and meat and are easy-to-handle pets, the perfect choice for inner-city dwellers.

But you don't have room? Don't think you could give them a good life? Well, think again. Consider the life of a commercially farmed chicken. They are given unnatural feed, little room to run around, and a shortened lifespan. I am sure any hen would prefer even a small back-yard with fresh food and room to move to a lifetime in a small metal cage under UV lights.

Fresh eggs from healthy hens are highly nutritious. The healthiest eggs come fresh from pasture-raised chickens that have been allowed to run around. Their droppings and henhouse straw are excellent fertilizers for the garden, and they can recycle your kitchen scraps, pick at your lawn, and rid your garden of bugs and grubs. What's more, they are gorgeous and hilarious creatures that will add character to any backyard or veggie patch.

Chickens are also wonderful pets for children. Some of my most precious childhood memories revolve around the chickens we kept in our backyard. I remember racing around the yard, and sometimes even through the house, determined to catch a hen or pat at one of our roosters' tails.

My chicken memories aren't entirely happy-go-lucky. We had one rooster that got so big and full of himself that he attacked my sister and ripped a large hole in her jeans. We were all petrified to go out to the backyard after that, but my clever mother somehow smuggled him out of the house in a box, took him to the Chinese woman up the street, and we all enjoyed a hearty meal of chicken soup the following evening.

Mom was thrifty enough not to be too sentimental about his demise. He'd had a good life and he came to a good end. My memory of what he looked like is a little vague, but I will never forget that soup—it was second to none.

But with the exception of the odd rowdy rooster, chickens make wonderful pets and often bond quite closely with humans. The year I broke up with a boyfriend, I moved to the outer suburbs of Melbourne and lived by myself with nothing but a dog and a hen to keep me company. It was a depressing year to say the least, but this darling bird—whom I'd bought from a nearby battery farm for less than the price of a hamburger—became one of my closest companions. She'd lay everyday—usually on the front doorstep, but sometimes in the house (she came in via the dog door, the clever thing). She particularly liked the cushions on the sofa, and was known to lay one beautiful egg in the center of the featherdown couch while I was out for the day. Her eggs were incredibly good, and they provided me with a wonderful sense of wholesomeness that I desperately needed at the time.

One day, while the front gate was open, poor little Egna ventured a little too far over our front fence and was bitten by the neighbor's dog. She had a broken wing and a pierced ribcage, and I knew there was only one option. I sped to the local vet, talking to her all the way, and had Egna put down. Had I been a bit more frugal, of course, I would have opted for the traditional nip and twist method used by farmers to finish off their hens. But I couldn't bear to. Hens are special creatures. So you've been warned: you may get attached. Despite her untimely death, Egna had six good months living with me—probably the best six months of her life. So despite my sadness at her passing, I was also very pleased that she had that period of freedom. Would her life have been any better or longer if I'd left her at the battery farm? I doubt it.

A GOOD EGG

The benefits of having your own hens go way beyond their beauty and friendship; they also produce glorious eggs and wonderful meat.

The best eggs come from hens that have had an omnivorous diet—lots of fresh greens, occasional kitchen scraps, and minimal grains—along with plenty of exercise. This is akin to their natural diet. Commercial chickens, however, are reared to produce more eggs and to yield meat as quickly as possible. They are fed an unnatural diet of grains, corn, or soymeal. Consequently, the chickens grow faster, and they produce more eggs—but the nutritional profile of their eggs and meat is different. Chickens fed a traditional diet and given room to run around develop healthier muscle meat. Their eggs and meat contain greater quantities of vitamins A, D, and E, as well as more omega-3 fatty acids, which are known to help prevent coronary-artery disease, hypertension, arthritis, cancer, and other inflammatory and autoimmune disorders. The yolks of their eggs will be a brighter orange and contain a higher level of health-promoting carotenoids. Their bones and cartilage contain more minerals, and they are not fed antibiotics or growth hormones.

By keeping your own hens or sourcing locally produced eggs, you can see what their diet and lifestyle are like, which is the best guarantee of good eggs. It will also mean that your eggs are fresh. I can't quite describe to you the beauty of fresh eggs. They are lighter, softer, and easier to poach. If you can't keep your own chickens, look for a local community garden that keeps chickens and has fresh eggs for sale.

If you can only buy eggs from a retail outlet, look for free-range *and* organic. There are sometimes loopholes in our systems of chicken certification. In some cases, hens can be certified as organic, but be fed a diet predominantly of grains. They can be labeled free-range but have relatively little room to run around in. So if you look for both

certification ticks, that is your best assurance of quality.

Raising your own chickens or sourcing them from a local farm also gives you access to quality meat, and to all the traditional bits. Chicken feet are considered the magic ingredient in chicken soup (also known as "Jewish penicillin"). They are incredibly cost-effective—one bag of chicken feet can make a large pot of wholesome and nourishing stock. There's also a certain feel-good factor in knowing that nothing has been wasted from the rooster or hen that ended up on the chopping block.

Choosing your chickens

Chickens come in all shapes, sizes, temperaments, and laying capacities. You can buy them from markets, wholesalers, or even at roadside stalls. Places that breed battery hens will often let you purchase their breeding hens for next to nothing when they are finished with them. Sometimes they give them away for free. These hens will be tame and quiet. You will get the satisfaction of giving them a second life and watching their personalities develop, as they go from being shy and timid to friendly and even a little boisterous.

If you are living in an inner-city or suburban area where noise is a problem, it is probably easiest to buy one of the more domesticated breeds such as Isa Brown or Australorp. These lay well, are very tame, and reasonably quiet. Silky Bantams (commonly referred to as "fluffy-bums") are also very good-natured, but tend to lay only a few eggs per week. Their eggs are well worth it, however—they are smaller than normal eggs and exquisitely delicious.

If you have a little more room, and more tolerant neighbors, you might try some of the heritage varieties of chickens such as Light Sussex, Leghorns, or Rhode Island Reds. The list is endless, and we should be supporting more of these rare breeds to keep diversity and flavor in our chicken market. They are a bit more expensive (you could be looking at forty or fifty dollars per hen) and their personalities may be less domesticated. But they are beautiful to look at, and you might just find yourself mooning over them most mornings, watching them rummage around your backyard.

What your hens will need

First, you'll need a secure henhouse that is impenetrable by foxes. If your chickens are out and about everyday, their henhouse doesn't need to be very big. For a while, when living in the inner-city, we had a henhouse that was 3 feet by 20 inches for two friendly Isa Brown hens. This was all the space they needed, as they only used the henhouse at night. You can buy henhouses online, or convert an old play house or shed. Hens like to be elevated, so if your henhouse isn't high off the ground, be sure it includes an elevated perch. This makes them feel protected from predators. They'll also need somewhere to lay, either nesting boxes or a pile of straw in their henhouse, and somewhere to shelter in extreme heat. You can insulate their henhouse with vines and straw. Chickens like to bury themselves in the dirt to keep cool, so let them do this.

Chickens need a constant supply of fresh water. Hens like their water cool (no warmer than 40°F) and they prefer a dripping tap or some other source of running water. It makes them feel like they are drinking from a stream rather than a stagnant pond. Adding a few drops of apple-cider vinegar to their drinking water every now and then will help to keep their digestive systems clean and free of parasites.

As omnivores, chickens like a diverse supply of food on demand. The easiest way to provide this is with a chicken feeder stocked with grains, supplemented by kitchen scraps. Chickens love their food fresh, especially their greens. When you are pruning your silverbeet or clipping your lettuces, be kind and throw some to your hens.

Clipping your chickens' wings (the feathers, obviously—not the bone!) will stop them from escaping. You need only clip one wing, and it isn't painful for them.

Very occasionally, you may find that your chickens have started pecking at their own eggs. This must be stopped immediately or they will make it a habit. You can buy fake eggs at pet shops. Put some of these in your henhouse and gather the real eggs as soon as they lay them. Within a few days they should lose interest.

CHICKEN FEED

Chickens love kitchen scraps, so this should be one of the first things you feed them. Chickens are omnivores (not vegetarians, as many people believe), so they require a varied diet of vegetables, grasses, worms, and some grains. They are also clever. They will pick through your kitchen scraps and take what they like. Very rarely will they eat anything that would make them ill. A tiny bit of meat in their diet is a good thing, but only feed it to them in moderation. In a natural environment they are always hunting for worms and snails, as these are the most nutritious food source. But in the wild, these are only available in small numbers, so form only a tiny part of their diet.

In addition to kitchen scraps, you should always have grains available for your hens, so that they can access food on demand. Various chicken feeders are available on the market, many of which are rat-proof and pigeon-proof. Wheat is the most commonly available chicken feed, but you can also feed them oats, maize, and sunflower seeds.

"Shellgrit," the leftover sea shells that wash up on the beach, is also essential in your chickens' diet, as it provides an important source of calcium to keep their bones and egg shells strong. You can collect it from the beach or purchase it from an animal feed store.

If you are buying chickens straight from a battery farm, you might find that they will only eat pellet feed for the first couple of days. Buy some of this pellet feed so that they can ease into their new diet. It may take them a while, but with time they should adjust to a natural diet of grasses, worms, and grubs.

SCRATCH THAT

Chickens love to scratch, and they love to peck and mow at grasses. They will keep your lawn nicely trimmed and eat any old lettuces and greens you have springing loose. However, be warned that if you let them loose in your vegetable patch they will get rid of any grubs and weeds, but they may also destroy some of your plants with their reckless pecking. Just be prepared!

An ideal set-up for chickens is to have them in a movable pen or within temporary fencing. That way you can rotate their position, moving them around your backyard so that they always have access to fresh grubs, dirt, and grasses. Another clever idea is to grow hardy fresh greens such as silverbeet or spinach around the outside of their henhouse. They'll always have a source of fresh greens, which they can peck at easily every day (ensuring very nutritious eggs), but they wont be able to destroy the whole plant.

TO ROOSTER OR NOT TO ROOSTER?

If you have the space, and the right council regulations, roosters are a huge amount of fun. They strut around the backyard, cavort with the hens, and make plenty of noise in the morning. But if you live anywhere in the inner city, roosters will not be a good idea. They do crow at sunrise and no amount of coaxing can prevent this (putting them in a small box at night can stop it, but I wouldn't advise this as a long-term solution!). In a farm environment, roosters protect the hens from predators during the daytime (nighttime is another story—they are useless) and establish a pecking order among the flock. They also make excellent chicken soup.

BEYOND THE CHICKENS: DUCKS, QUAIL, & GEESE

Chickens are the best food-providing animal for suburban environments. But if you have a bit more space and more distance from your neighbors, you can think beyond chickens. If you have room for a pond, you could consider getting a few ducks or geese. If you have room in your garden for a large cage, you could also consider getting some quail. Like chickens, ducks and quail provide wholesome, healthy eggs, as well as fertilizer for the garden. The drawback is that they can make more noise, and potentially a lot more mess.

SUPERBLY SCRAMBLED EGGS

For the best results, eggs need gentle cooking. This method of scrambling only *lightly* cooks them, so that they stay light and fluffy. They go deliciously well with caramelized onion and some crunchy sourdough toast.

Finely chop the onion. Put a little fat in a frying pan and place it over low heat. Cook the onion for 5 to 10 minutes, stirring intermittently, until it lightly caramelizes but does not burn.

While the onion cooks, whisk the eggs with a fork in a small bowl. Add salt and pepper to taste.

When the onion is done, transfer it to the bowl containing the eggs and wipe the frying pan clean.

Increase the temperature and add a small dollop of fresh fat to the pan. Pour in the egg and onion mixture.

Now it's time to scramble. The ideal technique is to constantly fold the egg mixture from the bottom of the pan to the top. This will keep it from overcooking and the middle section will stay beautifully soft and fluffy. Do this constantly for 1 or 2 minutes, or until the eggs are cooked to your liking.

Serve with some crunchy toast.

Preparation time: 5 minutes
Cooking time: 12 minutes
Serves: 1

Ingredients:
1 small onion
fat for frying
3 eggs
salt and pepper

EGG Mayonnaise

When I was growing up, egg mayonnaise was the ultimate "waste nothing" dish. Mom and I used to team up: I would make meringues using the egg whites and she would make mayonnaise using the yolks. Unfortunately, as a result, I was always hopeless at making mayonnaise, and she was rather bad at making meringues. This is therefore an extremely easy recipe with very few ingredients. If you have your own birds to provide the yolks, this mayonnaise is as cheap as chickens (excuse the pun).

Place the egg yolks, sea salt, vinegar, and lemon juice in a medium bowl and whisk well or beat with an electric beater.

Gradually, in a very thin stream, add the olive oil, whisking all the time. The whisking will be easier if the bowl is held at a slight angle, so you may want to make this a two-person job; one person can hold the bowl and whisk while the other gradually adds the oil in a steady stream. Or, if you're working solo, try this trick: layer two tea towels on the kitchen counter and prop the bowl against them, so that it sits at an angle. Your second hand will be free to pour the olive oil.

Preparation time: 4 minutes
Cooking time: none
Serves: 6 (as a condiment)

Ingredients:
2 egg yolks
1 teaspoon sea salt
1 or 2 teaspoons apple-cider or
   wine vinegar
juice of half a lemon
¾ cup olive oil
salt and pepper
to season, finely chopped fresh
   dill or 1 teaspoon Dijon
   mustard (optional)

Once all the oil has been added, taste the mayonnaise for acidity. If necessary, adjust by adding drops of lemon juice, salt, and pepper. If you are using extra flavorings such as herbs or mustard, add them now.

Variations:

AIOLI: Add 2 small cloves of crushed garlic to the egg yolks before you start whisking. Aioli is delicious drizzled on baked potatoes or with boiled eggs and salad greens.

SMOKED PAPRIKA MAYONNAISE: For every cup of mayonnaise, add 1 teaspoon of smoked paprika and a squeeze of lime juice at the end. This works well with poached eggs and bacon, or in a bacon and lettuce sandwich.

Tip:
If your mayonnaise curdles during preparation, don't despair: it can be saved! In a clean bowl, work an extra egg yolk into a smooth paste. Slowly add the failed mayonnaise, whisking well after each spoonful.

EGG & GREENS PIE

This pie is the perfect way to show off the latest produce from your backyard—against the green of the silverbeet, the orange egg yolks stand out like traffic lights. It's also a wonderfully simple dish, ideal for Sunday brunch or even, made the night before, for a weekday lunch. Whenever you serve it, it will definitely give you something to talk (or crow) about.

Preheat the oven to 350°F and grease a 10-inch pie tin.

Use your fingers to roll out the pastry into the pie tin. Bake the pastry blind (see note on page 253) in the oven for 15 minutes, or until the sides turn golden brown. Remove the bake-blind filling and return the pastry to the oven for another 5 minutes, just enough to make the bottom nice and crispy.

While the pastry bakes, finely chop the onion and gently cook it over low heat with a little fat. Let it simmer, stirring intermittently, for 5 to 10 minutes, or until it sweetens and turns golden. Don't let it brown.

Finely chop the silverbeet. Add it to the onion and let it simmer until it wilts. This usually takes only a minute or two. When the silverbeet has wilted, transfer the vegetables to your pie crust.

Preparation time: 20 minutes
Cooking time: 30 minutes
Serves: 4

Ingredients:
1 quantity of oatmeal pastry
   (see recipe on page 252)
1 small onion
fat for frying
2 or 3 large silverbeet leaves,
   stems removed
7 eggs
salt and pepper

Variation:
A few slices of ham or prosciutto go beautifully with the combination of silverbeet and onion.

Carefully break each egg into a glass and pour them onto the pie one by one (you don't want any broken yolks, as they won't look as striking). The recipe calls for 7 eggs, but for just the right balance of colors and flavors I usually use 5 whole eggs and 2 egg yolks, saving the 2 extra whites for other cooking.

Bake the pie for 20 minutes or until the eggs are set but still a little moist.

omelets

With all those chickens running around, you are going to have to think of creative ways to use up all the eggs! With a few fresh salad leaves and some crusty bread, a delicious, nourishing omelet can be ready in minutes. Omelets are a great way to celebrate whatever is currently fresh, whether it's a ripe zucchini or a tasty potato.

Omelets should be a special, simple delicacy—but things can go wrong. If the eggs overcook, they will be dry and tough. The best omelets strike a balance, so that the bottom is crispy but the middle and top are still delicately soft. You can achieve this by pan-frying first for a few minutes, then oven-baking. To do this you'll need a frying pan with an ovenproof handle, preferably one with a heavy base. I use a cast-iron Le Creuset pan that I bought from a thrift store—it dishes out superb, foolproof omelets. But don't worry if you haven't found your thrift-store Le Creuset yet: a basic stainless steel pan will be just fine. A small pan (about 8 inches in diameter) is best. If you're using a larger pan, you may need to adjust the quantities. The omelets described here are about an inch thick, crisp on the bottom, and soft on top.

ZUCCHINI & BASIL OMELET

Preheat the oven to its highest setting.

Slice the zucchini into thin rounds. Over medium heat, heat a teaspoon of fat and fry the zucchini until it is just tender.

Whisk the eggs, then add the basil and salt and pepper to taste.

When the zucchini is ready, transfer it to the egg mixture. Wipe any crumbs from the pan and return it to the stove and turn up the heat.

Add a small dollop of fat to the pan, then pour in the egg mixture and cook for 2 to 3 minutes, allowing the edges of the omelet to curl up.

Remove the pan from the stove and place it in the oven. Cook for 5 to 10 minutes, keeping a watchful eye on it to ensure that the eggs do not overcook. Remove the pan from the oven when the top of the omelet is set but still soft; it should be moist but not too runny.

Preparation time: 5 minutes
Cooking time: 10 minutes
Serves: 2

Ingredients:
1 small zucchini
fat for frying
3 eggs
1 handful of fresh basil, chopped
salt and pepper

POTATO & NUTMEG OMELET

Preheat the oven to its highest setting.

Wash the potatoes and chop them up into small cubes (about ⅓ or ¾ inch wide). Heat your frying pan over a medium heat. Add the fat, then fry the potatoes for 3 to 5 minutes or until they are lightly browned and cooked all the way through.

Whisk the eggs in a bowl with the salt, pepper, nutmeg, and sour cream.

Pour the cooked potatoes into the egg mixture. Wipe out the frying pan and return it to the heat. Let it heat up for about 30 seconds, then add some more fat.

Pour the egg and potato mixture into the pan, making sure the potatoes are evenly dispersed. Let the bottom and sides get crispy and the edges curl up.

Transfer the pan to the oven and bake for 5 to 10 minutes or until the top of the omelet is cooked to your liking.

Preparation time: 5 minutes
Cooking time: 10 minutes
Serves: 2

Ingredients:
2 or 3 small potatoes
fat for frying
4 eggs
salt and pepper
¼ teaspoon freshly ground
    nutmeg
¼ cup sour cream

Leek & Sour cream omelet

Preheat the oven to its highest setting.

Wash and finely slice the leek. Add some fat to the frying pan and sauté the leek over low heat for 5 minutes. It should soften but not brown.

Whisk the eggs with the sour cream and add salt and pepper to taste.

When the leeks are ready, add them to the eggs. Wipe the frying pan clean and return it to the heat. Let it heat up for about 30 seconds, then add some fresh fat and pour in the omelet mixture. Cook for 3 to 4 minutes or until the omelet curls up at the edges and lightly browns.

Remove the pan from the heat and transfer it to the oven. Cook for 5 to 10 minutes or until the top of the omelet is just cooked (or perhaps a little gooey, if you prefer).

Preparation time: 5 minutes
Cooking time: 10 minutes
Serves: 2

Ingredients:
1 medium leek
fat for frying
4 eggs
¼ cup sour cream
salt and pepper

Poultry Basics

Cooking with the whole bird is so much more economical than buying individual breast and thigh fillets. Not only do you get more meat for your money, you also get all the extra bits: gelatinous bones and feet for stock, and delicious poultry fat for home-cooking. These are some of the most nutritious parts of the bird, and are very handy to have on hand. To get the most from your bird, here are a few tricks you should have up your sleeve.

Slaughter: All this talk of chickens and cooking brings me to a tricky question: how to finish off your hen or rooster when the time comes. A generation or two ago, people slaughtered their chickens themselves or enlisted a neighbor. We were lucky when I was growing up; the infamous Chinese lady three doors down did the whole job—axe, pluck, and gut, all for a very small fee. But where is she now? Long gone, I suspect.

To finish off a bird, you can opt for either a traditional twist and pull or an axe and chopping block (the latter can be less confusing for the inexperienced chicken-slaughterer). It may sound confronting, but this process needn't be beyond suburban gardeners.

You'll find that your home-reared birds make wonderful chicken soup. I prefer to use older hens rather than the plumped-up youngsters, as they give the soup much more depth and flavor, and you have the satisfaction of knowing that the birds had long and happy lives.

Jointing a bird: Jointing a bird involves cutting up a whole bird into portions suitable for a crock-pot or casserole. Don't worry: it's easy! You'll need a sharp knife, a clean cutting board, and three bowls at the ready. As you work, use one bowl for the meat, one for the fat, and the other for bones and offcuts, which you can use for stock.

To start, place the bird on the cutting board, breast side up. If you are jointing a duck or goose, they have quite a lot of fat around the neck and bottom areas. Trim this off and place it in the fats bowl. Chickens have relatively little fat, but it's still worth keeping whatever fat you can find.

To start jointing your bird, first find the wing and cut off the wing tip. Put it in the offcuts bowl. If the bird still has its feet (this is unlikely; most butchers remove them), cut them off and add them to the offcuts bowl too.

Next, have a look for the neck. Your butcher may have removed it, but if it's still on the carcass you will need to remove it with the knife at the lowest point you can reach.

Now make a cut in the skin between one of the legs and the bird's body. Pull the leg towards you so that the thigh bone pops completely out of its socket. Cut the leg at the joint and remove it, placing it in the meat bowl. Repeat this process with the other leg, and then with the wings, popping the joints from the sockets before cutting them away from the carcass.

Slice along the backbone and remove the breasts, placing them in the meat bowl.

Lastly, trim any remaining meat from the carcass and place it in the meat bowl.

Congratulations: you've jointed a bird! You now have a whole animal's worth of meat, fat for frying or pastry-making, and a supply of bones and offcuts for stock.

If you're squeamish about jointing a chicken, you can still enjoy the benefits of purchasing a whole bird. Just ask your butcher if he's willing to "joint" the chicken for you. The hard work of snapping and pulling the bones will be done, and you'll just need to remove the meat.

To render the fat from your bird, see the recipes in the "Fats" chapter.

Giblets and gizzards: Giblets include the internal organs—the heart, liver, and any other thrifty bits. The gizzards include the stomach and intestines. The liver is best eaten immediately, pan-fried with some fresh herbs and enjoyed on toast. Alternatively it can be turned into pâté. Everything else can be added to the stockpot or tossed with some herbs and breadcrumbs and used as a stuffing if you are making a roast.

Reusing the bones from a bird after a meal:
When you've finished a meal of roast poultry, gather the bones—and I mean all the bones. It doesn't matter if people have held them in their hands or chewed on them; retain everything you can. The bones will be boiled at a very high temperature, so any bacteria will be killed off.

Put the bones in a stockpot, cover them with cold water, and add the head, neck, or wings if you have them. Throw in any vegetables or herbs you have on hand and make the stock according to the "Poultry Stock" recipe on page 128.

If you don't have time to make the stock right away, throw the carcass, plus any liquid that has accumulated at the bottom of the pan, the neck, and any other bits and pieces in the freezer, and make the stock whenever the mood takes you.

CHICKEN SOUP

This dish is a friend to come home to, a wholesome meal and a dose of Jewish penicillin all in one. One whole chicken can feed six people generously and will fill your kitchen with warmth and nourishment. Wherever possible, try to include the chicken feet when you make the stock. They add extra gelatin, which, according to folklore, is what gives chicken soup its famed medicinal powers. I think it's a real shame that when we buy a "whole chicken" from the butcher or supermarket we seldom get more than the body and the bones. The head and the feet are well worth using if you can get your hands on them.

Place the chicken and the vinegar in a large stockpot. Fill the pot with enough cold water to cover the bird. Bring the water to a boil, then reduce to a simmer. Chop up 2 of the carrots and 2 of the celery sticks and add them to the pot along with the leek, onion, peppercorns, kombu, and herbs. Simmer gently, partially covered, for 45 minutes or until the meat appears to be loosening from the bones.

Remove the chicken carcass from the pot, but leave the vegetables and any loose bones or feet to gently simmer.

Preparation time: 10 minutes
Cooking time: At least 1 hour
Serves: 4 (generously)

Ingredients:
1 whole chicken (including the head and feet if possible)
⅛ cup apple-cider or wine vinegar
4 medium carrots or parsnips (or a combination)
4 sticks celery, including the leaves
1 large leek
1 onion
1 heaped teaspoon black peppercorns
1 stick kombu seaweed (optional)
1 small handful fresh thyme and/or fresh rosemary
2 cups additional chicken stock (only if you opt for the shorter cooking time)

Using your hands, remove the meat from the carcass. Drain the meat of any excess fluid and place it in a small bowl. Sprinkle a layer of sea salt over the top and splash the chicken with olive oil. Put the bowl in the refrigerator.

Return the carcass to the cooking pot. You can make the soup now and it will be ready in 10 minutes. Or, you can let the stock continue to simmer, which will give it more richness and flavor. I sometimes let mine simmer for another hour or even, if I'm not in a hurry to use it, overnight. If you are going to make the soup right away, it can be a good idea to add some extra chicken stock for extra depth of flavor, but this is not essential.

When you're ready to make the soup, finely chop the remaining carrots and celery and the leek. Sauté them with a little cooking fat for 3 to 5 minutes or until they are lightly browned.

Transfer a few ladlefuls of stock from the stockpot into the saucepan with the vegetables. Simmer for a further 5 minutes, or until the vegetables are cooked.

Remove the chicken meat from the fridge and shred it into individual serving bowls. Ladle spoonfuls of the soup into each bowl and garnish with a generous drizzle of olive oil.

There is usually some leftover liquid, sometimes as much as 2 quarts of useful cooking stock. This can be stored in the fridge or freezer for later use.

Frugal Roast Chicken

Whether you buy your hen from your local butcher, a farmer, or the supermarket, as a frugavore you'll want to make the most of every last morsel. This classic roast chicken makes use of every last scrap, from the wings to the feet.

Preheat the oven to 350°F.

    If you have a whole chicken with head, neck, and feet still attached, chop these off and set them aside for stock-making. To remove the head, cut through the base of the neck where it reaches the top of the breast. If your chicken is already headless, you'll still need to remove the neck (it doesn't taste terribly good in a roast, so keep it for stock-making). Remove the "wingettes"—the end joint of the wings—by cutting at the joint and snapping them off (these can be kept for stock-making, too).

    Using your fingertips, poke the butter and a few snippets of the herbs into the space between the breast flesh and the skin.

    Cut the lemon in half and squeeze the juice over the whole bird. Rub the skin down with the remaining chopped herbs. Coarsely chop the onion and place half of it in the cavity of

Preparation time: 10 minutes
Cooking time: 1 hour
Serves: 4 to 6 (depending on the size of the chicken)

Ingredients:
1 whole chicken
2 teaspoons butter
1 small bunch thyme, finely chopped
1 small bunch rosemary, finely chopped
1 lemon
1 onion

the bird. Save the other half for stock-making, or for use in a side dish.

Put your bird, breast-side up, into a baking dish with a tight-fitting lid. The pot should be big enough to fit the bird snugly, without squashing it at the top or the sides; the lid will help to keep the meat moist. Place the dish, uncovered, in the oven and roast for 15 to 20 minutes or until the bird is nicely browned.

Remove the dish from the oven and turn the chicken over so that it now faces breast downwards. Add enough water to the dish to fill it ¾ to ¼ inches deep, then put the lid on. Return the dish to the oven and cook for a further 45 minutes. To test whether the bird is ready, poke one of the drumsticks with a skewer. When the bird is ready the juices will be clear, not pink.

Don't forget that at the end of the meal, the scraps, carcass, and offcuts can all be used for stock, while any excess fat can be rendered to make cooking fat.

ROAST DUCK WITH orange & sage

Duck has long been prized for its rich, dark flesh and delicious fat. Although ducks are not cheap, they can still be economical, as a single duck contains ingredients that can be put to good use in multiple dishes. Duck fat is perfect for cake-making and frying, and duck bones make a beautiful stock (lentil soup with duck stock is a favorite at our house).

Preheat your oven to 425°F.

Take a look at your bird. If your duck has come with a head, neck, or feet, you need to remove these by snapping them and then cutting them off with a sharp knife. Don't throw them away; stash them in your fridge or freezer for the next time you make stock. Trim off any excess fat from the neck of the duck. Use a spoonful for roasting vegetables (see side note) or save it to be rendered for cooking fat.

Rub the duck down with salt and pepper to taste. To make a thrifty stuffing, cut a few slices of onion and orange and place these in the cavity of the duck. Some breadcrumbs and herbs also work well. If your duck has come with giblets or gizzards, you can finely chop them and add them to the stuffing too.

Preparation time: 15 minutes
Cooking time: 1 hour and 20 minutes
Serves: 5 (approximately, depending on the size of the duck)

Ingredients:
1 whole duck
1 slice sourdough bread
1 medium onion
1 small handful fresh sage, finely chopped
1 small handful fresh thyme, finely chopped
olive oil
2 oranges

Truss the duck by tying together the legs with a piece of string to keep them close to the body. With a sharp-tipped knife or fine needle, carefully poke a few holes on the surface of the duck to allow the fat to permeate throughout the skin. Squeeze half of one orange over the duck and rub thoroughly.

Place the duck on its side in a baking dish. If necessary, secure it in place by chopping an onion in half and placing the pieces on either side of the bird. Roast the duck for 20 minutes, then turn it over and cook for a further 20 minutes on the other side.

Remove the tray from the oven and turn the duck breast-up. Slice the orange very thinly and place the slices on the up-turned breast of the bird. Reduce the oven temperature to 400°F and return the duck to the oven. Cook for a further 20 to 30 minutes or until the skin is crispy. Duck is properly cooked when the temperature of the meat in the thickest part of the thigh or breast reaches 165°F (this can be checked with a meat thermometer). Alternatively, just look for crispy skin and pink juices from the thickest portion of meat. When the duck is ready, transfer it to a serving tray and enjoy with roasted vegetables.

Note:
To prepare a side dish of vegetables, arrange an assortment of pumpkins, carrots, and potatoes in a separate baking tray. Cut off ¾ teaspoon of fat from the rear of the bird and add this to the tray along with some fresh sage and thyme. Bake in the oven alongside the duck for 50 to 60 minutes or until golden and crispy.

spanish-style chicken casserole

This recipe uses up all the bits—the bones can be set aside for stock, the meat goes into the casserole, and the fat can be rendered for later cooking. It's also an excellent way to use up any extra tomatoes that are lurking in the back of the fridge (soft and squishy is fine for this recipe). Serve with roasted potatoes or toasted polenta and a fresh garden salad.

Preheat the oven to 350°F.

If your chicken has come with feet, a neck, or "wingettes" (the bone at the end of the wing), remove these and set them aside to use in stock. Trim off any excess fat and save it to be rendered into cooking fat. Joint the chicken, cut the meat into casserole-sized chunks, and put the carcass aside for stock.

Have ready a heavy-based cooking pot with a tight-fitting lid. Peel a long piece of lemon rind and add this to the pot along with the tomatoes, onion, stock, bay leaves, paprika, cayenne pepper, and garlic. Place the pot on the stove and bring the stock to a gentle simmer.

In a frying pan over a high heat, fry each piece of chicken with a little cooking fat for about a minute on each side to seal

Preparation time: 5 minutes
Cooking time: 1 hour
Serves: 4 (depending on the size of the chicken)

Ingredients:
1 large free-range chicken
rind of 1 lemon
2 cups soft tomatoes (or 1 can preserved tomatoes)
1 large onion, finely chopped
2 cups stock
2 bay leaves
½ teaspoon paprika
¼ teaspoon cayenne pepper
6 large cloves garlic, crushed
fat for frying
½ cup pitted olives

in the flavor, then transfer them to the cooking pot, ensuring that there is enough liquid to cover the meat.

Bake in the preheated oven for 40 minutes. To test if the chicken is ready, poke a skewer through one of the drumsticks. When the meat is no longer pink, remove the pot from the oven and transfer the pieces of chicken to a serving tray.

Return the cooking pot to the stovetop. Add the olives, then bring the liquid to a boil over high heat and simmer for a good 10 minutes to allow the sauce to thicken.

Season the sauce to taste, then pour it over the roasted chicken and serve.

CHICKEN & LEEK PIE

There is nothing quite as special as the combination of chicken, leeks, tarragon, and dill. If you grow the herbs yourself, the flavor will be all the richer for it. I use a whole chicken for this dish, as it's significantly cheaper to buy the whole bird. However, you can make this recipe using pre-cut chicken breasts or thighs—a less frugal approach, but quicker. I've included both options in this recipe.

First, prepare the pastry and preheat your oven to 350°F.

If you are using a whole chicken, cut off any excess fat and debone the bird according to the instructions on page 108. Cut the meat into bite-sized chunks. Alternatively, if you are using ready-cut thighs and breasts, cut them into bite-sized pieces.

Heat a saucepan over medium heat, add a little fat and fry the chicken pieces in batches for 30 to 60 seconds on each side. They should seal but not cook through. Transfer them to a separate dish and wipe the pan clean.

Slice the mushrooms into ¼-inch pieces. Fry these in the pan with some fat for a couple of minutes or until they lose much of their moisture. Set them aside with the chicken pieces and again wipe the pan clean.

Preparation time: 1 hour
Cooking time: 1 hour
Serves: 6

Ingredients:

1 quantity oatmeal pastry (see recipe on page 252)
1 whole chicken (about 3⅓pounds), or 1¾ pounds chicken breast or thigh
fat for frying
½ pound mushrooms
2 leeks
1 cup sour cream
2 eggs
¾ tablespoon finely chopped tarragon
½ tablespoon finely chopped dill
1 teaspoon mustard
salt and pepper
¼ cup chicken stock or water
3 teaspoons arrowroot powder

Finely chop the leeks, then fry them gently with a little fat, stirring occasionally, for 10 to 15 minutes or until they are soft but not browned.

While the leeks cook, prepare the sauce by combining the sour cream, 1 of the eggs, and the tarragon, dill, and mustard. Season with salt and pepper to taste, then stir well and set aside.

In a small saucepan, heat the stock until it is warm but not boiling. If you don't have any stock, use the equivalent quantity of water. When the liquid is warm, add the arrowroot powder and stir to produce a smooth paste.

To the pan containing the cooked leeks, add the chicken, mushrooms, sour-cream mixture, and arrowroot paste. Simmer gently, stirring regularly, for 5 minutes. The mixture should be thick and gluey; if it's not, boil it down further and add some more arrowroot powder. When it is ready, transfer the filling to a large casserole dish.

Roll out the pastry to form a lid the size of your casserole dish and place it over the chicken filling. Prepare a glaze by whisking the remaining egg in a small bowl and brushing it over the pastry.

Place the pie in the oven and cook for 45 minutes, or until the crust is golden and crispy and the chicken has cooked through. Serve with a crispy green salad or cooked vegetables.

STOCKS & SOUPS

"To make a good soup, the pot must only simmer, or 'smile.'"

—French proverb

SOUP HAS ALWAYS BEEN A MAINSTAY OF PEASANT fare, the perfect way to combine the cheapest ingredients or even scraps (bones, knuckles, some chicken feet) to make a delicious and nourishing broth. A good pot of soup should restore and nourish. In France, the word *restoratif* described soups and stews sold at roadside taverns to weary travelers. They were a source of sustenance for those who were unwell and an antidote to physical exhaustion. The concept of a *restoratif* bowl of soup eventually expanded to describe any place people could stop for a sustaining meal—that is, a restaurant.

Stock is the basis of all good soup-making. Stock can be made from almost anything—bones, heads, tails, vegetable scraps, and cooking water. Making stock out of odds and ends reduces the cost of home-cooking and also decreases kitchen waste. Bone stocks made using the gelatinous parts of the animal are the most nutritious. Many traditional groups placed enormous value on bone stock. It was inexpensive and easy to make, was an important means of keeping well nourished, and ensured that no part of the animal went to waste. Gelatin-rich stock has many powerful properties. It can strengthen the skin, cartilage, bones, heart, muscles, and immune system. Animal feet, marrow, and shank are particularly rich in gelatin, so don't throw those chicken feet out—use them! Oxtail is also exceptionally good and is often very cheap, as not many people seem to buy them.

In addition to animal sources, you can also use vegetable scraps, the juice from lentils, beans, and legumes, or the cooking water left over after boiling vegetables. These do not have the same powerful properties as bone broth, but will still add flavor and nutrients to soups, stews, and casseroles.

Buying stock bones

When you are shopping, make sure you ask your butcher for "stock bones." If he or she doesn't understand what you're after, explain that you need cartilage and gelatin, for instance shank, marrow, or feet. If your butcher is a friendly one, request that the bones be chopped into smaller bits; this will make cooking them all the easier.

From a price perspective, bones are a good point to haggle on. Make sure you ask for anything they won't be needing and you'll most likely get a good price. You can also ask for "dog bones" or "scraps," as they are essentially the same thing.

Making stock

Making stock is all about extracting the nutrients and flavors from animal bones or vegetable scraps so that you can use them in later cooking. You can use any odds and ends: heads, tails, feet, and other offcuts such as shank, marrow, and rib. These offcuts are usually sold at a reduced price and are sometimes given away for free.

The golden rule when making stock with animal bones is to adjust the cooking time to suit the size of the bones. Fish bones are fairly small and thin, so you only need to cook them for two or three hours. Rabbit, chicken, and poultry bones are slightly bigger, so stock using these ingredients can be gently simmered for eight to ten hours. Beef and lamb bones take the longest and should be cooked for at least twenty hours, or overnight.

In all honesty, all you need for a good basic stock are some bones, water, and a small amount of vinegar. The other ingredients in the following recipes are really optional extras. They make the stock delicious, but don't go busting your chops if you don't have them. It's definitely not worth a trip to the supermarket: just make do with what you've got. And don't forget, stock-making is a great way to use leftovers. Stale and floppy carrots, that tired-looking celery that's sitting at the back of your fridge—these are all perfect ingredients for stock. You definitely don't need vegetables that are super fresh, as after a couple of hours on a gentle simmer, you won't know the difference anyway.

Whatever ingredients you're using, coarsely chop them and place them in a large stockpot with cold water. Bring to a boil, then simmer, partially covered, for several hours. Drain the juice and discard the solids, then refrigerate or freeze your stock in an airtight container. If you like, you can place the stock in the fridge, allow a layer of fat to form, then skim the fat from the surface and discard it.

STORING YOUR STOCK

Stock will keep in the freezer for months and is incredibly handy for last-minute dinners. If you store it in meal-sized portions, you can defrost them as you need them. It can also be boiled down to a thick, jelly-like consistency so that it takes up less space in your fridge or freezer. I store mine in recycled plastic ice-cream and yogurt containers. Alternatively, you can use zip-lock bags. Some people prefer to use glass jars. If using glass, only fill the jar three-quarters full, as the liquid will expand when it freezes.

CHICKEN OR POULTRY STOCK

If your butcher is willing to sell them to you, chicken heads, feet, and wings make excellent additions to stock. If you are buying your bird directly from the farm, it shouldn't be a problem to get the feet as well as the whole bird. If you can't get the feet, ask your butcher for the wingettes, which are also rich in gelatine and cartilage. This recipe is a master recipe for chicken stock, but it can be made using any poultry bird such as duck, goose, turkey, or pheasant. Larger birds with bigger bones require a longer cooking time and a few extra herbs.

Roughly chop the vegetables and crack the bones. Place all the ingredients in a large pot and cover them with cold water. Bring the stock to a boil, then allow it to simmer, partially covered, for 5 to 12 hours (if you are using a larger bird such as turkey or goose, let it simmer for a bit longer). When you've finished cooking, drain and discard the solids and retain the liquid.

Preparation time: 10 minutes
Cooking time: at least 5 hours
Makes: 1 gallon

Ingredients:
4½ pounds chicken carcasses and/or offcuts such as wingettes or feet
1 or 2 onions
2 or 3 carrots, parsnips, or turnips (or a combination)
2 sticks celery, including the leaves
1 bunch fresh thyme or rosemary
1 gallon cold water, or enough to cover all your bones
10 peppercorns
½ cup apple-cider or wine vinegar

Optional:
2 large fennel
1 or 2 leeks
A knob of fresh ginger

CHICKEN-FEET STOCK

Chicken feet are the magic ingredient in this rich and beautiful stock. They are full of gelatin, which imparts a special quality to soups and other dishes. If you are shopping directly from a farm or a knowledgeable butcher, you should be able to buy chicken feet in bulk—many good butchers sell them by the bag for less than the cost of a whole chicken. Making stock from chicken feet is very similar to standard chicken stock, but a bit of extra spice adds extra warmth.

Put the chicken feet into a large pot and cover them with cold water. Add the vinegar and slowly bring to a simmer.

Meanwhile, chop the vegetables and the herbs. When the water is simmering, add them to the pot.

Simmer, partially covered, for at least 5 hours. Check the water levels from time to time; if the chicken feet are not covered with water, add a little more. When you've finished cooking, drain and discard the solids and retain the liquid.

Preparation time: 10 minutes
Cooking time: at least 5 hours
Makes: 1⅓ gallons

Ingredients:

4⅖ pounds chicken feet
½ cup apple-cider or wine vinegar
1 large onion
2 or 3 carrots, parsnips, or turnips (or a combination)
3 or 4 sticks celery, including the leaves
1 bunch fresh thyme or rosemary
1 heaped teaspoon black peppercorns
a few slices of fresh ginger

Optional:
2 large fennel
flat-leaf parsley, added 10 minutes before you finish
nettle herbs or roots (dried or fresh)
a pinch of cayenne pepper

beef, lamb, or pork stock

Crack the meat bones and roughly chop all the vegetables. Place the bones and the vinegar in a large pot. Cover them with cold water and slowly bring to a boil. Add the herbs, peppercorns, and water then simmer, partially covered. The stock should not be boiling; it only needs to simmer or "smile."

You can cook your stock for anywhere between 3 and 48 hours; the longer, the better. As it cooks, skim off and discard any residue that accumulates on the surface.

When you have finished cooking, drain and discard the solids and retain the liquid. If you'd like to reduce the amount of fat, place the stock in the fridge, allow a layer of fat to form, then skim it off and discard it.

Preparation time: 10 minutes
Cooking time: at least 3 hours
Makes: 1⅓ gallons

Ingredients:
4⅖ pounds soup bones
2 large carrots, parsnip, or turnips (or a combination)
2 stalks celery, including the leaves
1 or 2 onions
¼ cup apple-cider or wine vinegar
1 bunch fresh thyme or rosemary (or a combination)
10 black peppercorns or 1 teaspoon black pepper
1 gallon cold water

Optional:
nettle herbs or roots (dried or fresh)
dried arame or kombu seaweed
flat-leaf parsley, added 10 minutes before you finish

VEGETABLE STOCK

This is really an exercise in cleaning out your fridge or freezer. You can add whatever you like, in whatever quantity suits you. The recipe below is just a guide.

Coarsely chop all the ingredients and place them in a large pot. Cover with cold water. Bring to a boil, then reduce the heat and simmer, partially covered, for 1 to 3 hours. When you've finished cooking, drain the juice and discard the solids.

Note:
The water left over after cooking chickpeas or beans can also be used as a vegetable stock. After cooking your beans, retain the juice and use it as a stock in casseroles, soups, or sauces. Bean juice does not contain all of the nutrients found in bone-based stock, but it does add flavor to any dish and is rich in vegetable-based nutrients.

Preparation time: 10 minutes
Cooking time: 1 to 3 hours
Makes: 2 quarts

Ingredients:
1 onion
2 leeks
3 stalks celery, including the leaves
4 parsnips or carrots
1 turnip
¾ inch ginger
10 black peppercorns or 1 teaspoon black pepper
1 teaspoon sea salt
8½ cups water
1 bouquet garni

Optional:
2 large fennel
flat-leaf parsley (added 10 minutes before you finish)
nettle herbs or roots (dried or fresh)
dried arame or kombu seaweed flakes

STOCK-BROTH COLD CURE

This concoction will defeat any cold, flu, or cooties that might be plaguing your household. Cheaper than acetaminophen and quicker than antibiotics, it will have the bugs literally flying out from under your nose. You can use any stock for this recipe, but I find that a rich beef stock works best. Feel free to add more garlic if you wish, and try not to cook the garlic too much, if at all.

Bring the stock and the water to a boil. Reduce the heat and add the garlic and sea salt. Serve in a large mug. After drinking it, have a rest for 20 minutes. On awakening, your cold should be gone, or your money back …

Preparation time: 2 minutes
Cooking time: 3 minutes
Serves: 1

Ingredients:
1½ cups strong stock
½ cup water
3 large cloves garlic, crushed
    and finely chopped
2 teaspoons sea salt

french onion soup

This is a very handy recipe if you don't have many fresh ingredients on hand. All you need is some stock, onions, wine, and a little bit of crusty bread and cheese. Gruyere is traditional, but a quality parmesan or mozzarella can be used instead.

Preheat your oven to 390°F, using the grill setting.

Finely slice the onions. In a large frying pan, heat a teaspoon of fat over low heat. Add the onions and cook them gently for 20 to 30 minutes, stirring occasionally. The onions should caramelize, turning a gentle golden color, but should not brown or burn. When they are almost ready, crush the garlic, chop it finely, and add it to the pan.

In a large pot, bring the stock to a boil, then add the white wine, brandy, and onions. Simmer gently for 25 minutes.

Thickly slice the bread and lightly toast the slices in the toaster.

Pour the soup into individual soup bowls. Garnish each bowl with a few slices of toast and a generous sprinkling of cheese. Season to taste, then place the bowls in the preheated oven for 5 to 10 minutes or until the cheese has melted.

Preparation time: 10 minutes
Cooking time: 1 hour, 10 minutes
Serves: 4

Ingredients:
4 or 5 medium onions
fat for frying
2 cloves garlic
2 quarts beef stock
½ cup white wine
⅛ cup brandy (optional)
a few pieces of crusty bread
5⅓ ounces Gruyere, parmesan, or mozzarella cheese, grated
salt and pepper

pea & ham soup

You can use either fresh or smoked ham bone (or hock) for this dish. Smoked ham has a wonderfully charismatic odor as it simmers on the stove. We usually use the leftover ham from Christmas dinner—a good pot can last us several days, which means investing in a healthy, free-range pig is well worth it.

Finely chop the onion. Dice the carrots and celery and crush and finely chop the garlic.

Place your ham bone in a large pot and completely cover it with water. Add the split peas, vinegar, bay leaf, thyme, and onion and bring to a boil. Reduce the heat and simmer gently for about 90 minutes, or until the meat begins to fall off the bone. Add the carrots, celery, and garlic and simmer for a further 30 minutes.

When the vegetables are soft and well cooked, remove them from the pot and place them in a food processor. Purée them until they are smooth, then return them to the pot. Alternatively, you can keep everything in the pot but push the meat to one side and use a hand-held blender to purée the vegetables.

Pull the meat from the bone and stir it through the soup.

Preparation time: 10 minutes
Cooking time: 2 hours
Serves: 8

Ingredients:
1 large onion
3 large carrots
3 large sticks celery, including the leaves
3 cloves garlic
1 ham bone, with or without meat (roughly 1¾ pounds)
2 cups green or yellow split peas
⅛ cup apple-cider or wine vinegar
1 bay leaf
2 teaspoons fresh thyme, finely chopped
1 handful fresh parsley, finely chopped
1 handful fresh mint, finely chopped

If necessary, adjust the water levels by adding a little more or by boiling down the stock. When you are happy with how it's looking, add salt and pepper to taste.

Pour the soup into individual bowls and sprinkle generously with freshly chopped parsley and mint. Drizzle with olive oil and serve with crusty bread.

minestrone

Nothing beats a good minestrone, with its rich stock, fresh vegetables, and nourishing legumes. Your minestrone will reflect what you have on hand in your pantry or back garden. Don't be afraid to play around with the ingredients, swapping carrots for parsnip or onions for shallots if it suits you. And if you happen to have something fresh and exciting from the garden, be it crunchy green beans, an unusual variety of turnip, or a handful of spinach, don't be afraid to throw it in as well.

Rinse your pre-soaked beans with cold running water. Place them in a large soup pot and cover generously with water. Finely chop the onion. Add the onion, fresh herbs, and stock to the pot and bring to a boil. Reduce the heat and simmer, partially covered, for 45 minutes or until the beans are soft.

Cut the carrots, green beans, leek, and potatoes into bite-sized pieces. Add them and the tomatoes to the pot and simmer for a further 15 minutes or until the vegetables are soft.

Preparation time: 10 minutes
Cooking time: 1 hour
Makes: 1 gallon (serves 8 to 10)

Ingredients:
1 cup chickpeas (dry weight),
    soaked overnight
½ cup white beans (dry weight),
    such as cannellini or haricot,
    soaked overnight
1 large onion
1 small handful fresh rosemary,
    finely chopped
1 small handful fresh thyme,
    finely chopped
1 quart rich stock (I find that
    chicken stock works best)
3 medium carrots
1 cup green beans
1 leek
3 medium potatoes
1 can diced or whole tomatoes,
    or 4 large and juicy fresh
    tomatoes

Ingredients continued ...

2 cups fresh greens such as
    silverbeet, cavolo nero, or
    spinach
2 cups shredded cabbage
salt and pepper
olive oil
freshly grated parmesan

Finely chop the greens and finely shred the cabbage. Add them to the pot and cook them for just a few minutes. Season your soup to taste, then pour it into individual serving bowls. Drizzle each bowl with some extra-virgin olive oil and sprinkle with parmesan cheese.

POTATO & LEEK SOUP

Two simple ingredients, a few dried herbs, dinner ready in thirty minutes. Enough said.

Slice the leeks into pieces about half an inch thick.

In a large pot, heat a teaspoon of fat over low heat. Add the leeks and half a cup of water. Cover and simmer gently for 5 minutes with the lid on, then 10 minutes without the lid. Allow the leeks to soften and sweat, but do not let them brown.

Meanwhile, coarsely slice the potatoes. When the leeks are ready, add the potatoes to the pot along with the thyme, nutmeg, chicken stock, and two cups of water. Bring to a boil then simmer, partially covered, for 10 minutes, or until the potatoes are soft.

Transfer ¾ of the soup to a blender and purée it, then return it to the pot. Alternatively, leave all the soup in the pot and purée it with a hand-held blender, leaving a few chunky bits.

Season to taste and serve.

Preparation time: 10 minutes
Cooking time: 15 minutes
Serves: 5

Ingredients:
3 large leeks
fat for frying
1½ pounds potatoes
2 teaspoons fresh thyme
1 teaspoon dried nutmeg,
    freshly grated
1½ quarts chicken stock
salt and pepper
olive oil

To serve:
I like a generous dollop of Greek yogurt or sour cream, some fresh parsley, and a good drizzle of olive oil. My partner recommends some fried bacon, cut into small pieces and sprinkled over each bowl.

CHICKEN & CORN SOUP

This recipe is a godsend for busy nine-to-fivers. If you have a whole chicken, simply slice off the breast for this recipe and use the rest of the bird for chicken soup or chicken and leek pie.

In a small saucepan, bring the chicken stock to a boil. Reduce the heat until it is gently simmering, then add the whole corn cobs. Let them cook for 10 minutes or until they are soft.

Remove the corn from the stock and scrape the kernels from the cobs. Place the corn kernels in a food processor with $\frac{1}{3}$ cup of the stock and all of the arrowroot powder. Purée to a fine paste, then return the mixture to the pot and stir well. Season to taste. The soup should now be thick and gluey. If it doesn't seem thick enough, add a little more arrowroot powder.

Finely slice the chicken meat and add it to the pot. Reduce the temperature and cook gently until the chicken is cooked through. Remove it from the heat and add the vinegar and soy sauce. Stir well. Serve with sprinklings of finely chopped spring onion.

Preparation time: 10 minutes
Cooking time: 20 minutes
Serves: 2

Ingredients:
1$\frac{2}{5}$ quarts chicken stock
2 fresh corn cobs
2½ tablespoons arrowroot
    powder
1 medium chicken breast or
    thigh, skinned and boned
1 teaspoon umeboshi plum
    vinegar
½ teaspoon soy sauce
2 spring onions

saffron stracciatella

Remember all those tupper ware containers filled with chicken stock that are sitting in your freezer? Well, now is the time to use them up. Even if you are exhausted, hungover, or only have one functioning hand, you can make this soup. It's like a fluffy rug and hotwater bottle, only tastier.

In a medium saucepan, bring the chicken stock to a gentle simmer. Add the saffron threads and stir well.

In a small bowl, combine the eggs, cheese, parsley, breadcrumbs, and salt and pepper to taste. Whisk until the eggs are well combined.

In a steady stream, pour half of the chicken stock into the bowl, whisking as you pour. Whisk well for a further 30 seconds, then pour the mixture back into the saucepan. Return to the boil and cook for 2 minutes, gently whisking all the time. The eggs and breadcrumbs should clump together.

Voilà. Serve, garnished with parmesan cheese and a dash of olive oil.

Preparation time: 5 minutes
Cooking time: 5 minutes
Serves: 2

Ingredients:
1 quart chicken stock
½ teaspoon saffron threads
3 eggs
¼ cup finely grated parmesan, plus extra for serving
¼ cup flat-leaf parsley, finely chopped
1 slice of bread, crusts removed, ripped into small breadcrumbs
salt and pepper
olive oil

CHICKPea & ROSeMary SOUP

This soup is best made with ripe tomatoes and home-grown rosemary freshly plucked from your backyard pots. I have on occasion resorted to stealing rosemary from over a neighbor's fence. But the soup is so good, this little theft seems worth it; just think of it as positive pilfering ...

Soak your chickpeas overnight.

Finely chop the onions and dice the potatoes.

Heat a little butter in a large pot and gently fry the onions for about 5 minutes, or until they are soft and brown. Finely chop the garlic and add it to the pot, along with the rosemary leaves. Turn down the heat and simmer for 3 minutes.

Drain the chickpeas and discard the soaking water. Add the chickpeas to the pot along with the tomatoes, water, and stock.

Bring to a boil and simmer for one hour, or until the chickpeas are soft. When they are beginning to soften, add the potatoes and stir well. Cook for a further 10 minutes or until the potatoes are cooked.

Season to taste and serve with a generous sprinkling of finely chopped flat-leaf parsley and a drizzle of olive oil.

Soaking time: overnight
Preparation time: 10 minutes
Cooking time: 1 hour
Serves: 8

Ingredients:
2 cups chickpeas (dry weight), soaked overnight
3 onions, finely chopped
4 medium potatoes
butter for frying
2 garlic cloves, crushed and finely chopped
3 large stems of rosemary, about the length of your hand
8 ripe tomatoes, roughly chopped, or 2 cans of pre-served tomatoes (fresh is better)
1 quart water
1 quart chicken stock, vegetable stock, or leftover bean juice
salt and pepper
1 large handful flat-leaf parsley
olive oil

PUMPKIN SOUP

Pumpkins manage to pop up in all sorts of peculiar places in our garden. Because all of our vegetable scraps are composted and ultimately end up back on the veggie patch as fertilizer, it's no surprise that little pumpkin sprouts seem to appear here, there, and everywhere when spring comes. If you have a wide open space, pumpkins can go a bit bananas; this is an excellent way to use them up.

Uncut and unopened, pumpkins can sit happily in a cool dark place such as a pantry or cupboard for weeks, if not months. Store them there until you feel like making soup. They also look fabulous stacked on your kitchen table or mantelpiece.

Chicken stock is traditionally used as a base for pumpkin soup, but I have also had success with lamb stock. They make for a soup that is richer and heavier, but still delicious.

Preparation time: 10 minutes
Cooking time: 30 minutes
Serves: 8

Ingredients:
1 large onion
fat for frying
1⅓ pounds Japla pumpkin (roughly half of one large pumpkin)
1 pound butternut pumpkin (roughly half of one large pumpkin)
2 medium carrots
1 quart chicken stock
2 teaspoons allspice (or 1 teaspoon dried cinnamon and 1 teaspoon nutmeg)
½ teaspoon dried ginger
1 clove garlic
to serve, sour cream or yogurt, and fresh parsley or chives

Coarsely chop the onion. Heat a little fat in a soup pot over a medium heat and sauté the onion for 5 to 10 minutes or until it turns a rich golden color.

Meanwhile, coarsely chop the other vegetables. When the onion is ready, add the vegetables, stock, spices, and garlic to the pot, plus a little extra water if necessary to completely immerse everything. Bring to a boil, then reduce the heat and simmer for 20 minutes, or until the vegetables are soft.

Purée the vegetables, either with a handheld blender or in a food processor.

Season to taste, then serve with dollops of sour cream or yogurt and a sprinkling of freshly chopped parsley or chives.

KaLe & zucchini soup

This is an excellent mid-week pick-me-up when you are feeling fried and tired after one too many deli coffees. In one
bowl of soup you will be getting about five whole vegetables.
The seaweed adds extra minerals and a subtle salty
flavor.

Slice the carrot into 1-inch pieces and cut the zucchini in half.
Rip the leafy greens into large chunks. Place the vegetables in
a small saucepan with just enough water to cover them. Add
the seaweed, then bring to a boil and simmer for 5 minutes or
until the vegetables are soft.

Transfer the mixture to a blender or food processor and
process until it forms a fine purée; it should be thick, like
heavy cream. As it is blending, add the basil and coriander
and season to taste.

Serve with a dollop of plain yogurt and a dash of olive oil
for each bowl.

Preparation time: 2 minutes
Cooking time: 10 minutes
Serves: 2

Ingredients:
2 small carrots
3 medium zucchini
4 large leaves of dark leafy
    greens (kale or cavolo nero),
    stems removed
2 strips kombu seaweed
¼ cup finely chopped basil
¼ cup finely chopped coriander
a few generous dollops of plain
    yogurt to serve
sea salt
olive oil

LENTIL SOUP

Lentil soup is the ultimate frugavore food. Lentils and stock bones, both wonderfully cheap, form the hearty base, complemented by whatever vegetables are fresh in your garden or at the market. Enjoy this soup as a mid-week meal, or take it to work for lunch in a thermos.

Finely chop the onion, garlic, and celery, and dice the zucchini and carrots.

Place the lentils and the stock in a large soup pot over medium heat. Add the garlic, onions, thyme, bay leaf, and tomatoes and simmer, partially covered, for about 30 minutes.

Once the lentils begin to soften, add the carrots and zucchini and cook for a further 10 or 15 minutes.

When the lentils and vegetables are cooked right through, season to taste with salt and pepper. Add the celery and chopped greens and simmer for a further minute or so.

Serve with a generous sprinkling of flat-leaf parsley and a drizzle of olive oil for each bowl.

Preparation time: 10 minutes
Cooking time: 1 hour
Serves: 5

Ingredients:
1 large onion
2 cloves garlic
2 sticks celery, including the
    leaves
1 medium zucchini
2 medium carrots
1½ cups dark red lentils
1½ quarts stock
2 teaspoons thyme leaves,
    fresh or dried
1 bay leaf
4 large ripe tomatoes, or
    1 can (14 ounces) whole
    peeled tomatoes
1 cup chopped fresh greens
    such as cavolo nero, silver-
    beet, or spinach
1 handful fresh flat-leaf
    parsley
salt and pepper
olive oil

Bean & Green Soup

Silverbeet is one of the easiest vegetables to grow yourself. With some fresh leaves and a few staples from your pantry, you can have this nourishing soup ready in less than an hour.

Finely chop the onions and garlic. Drain the beans and discard the soaking water. Place them in a large pot with the stock, bay leaf, onions, cumin, paprika, and garlic. Add two cups of water and bring to a boil. Reduce the heat and simmer, partially covered, for 40 to 50 minutes, or until the beans soften.

    Chop the silverbeet into bite-sized pieces and add these to the pot. Cook for 5 to 7 minutes, or until the leaves are soft and wilted. Season to taste, then serve with a dollop of ricotta cheese and a dash of olive oil.

Soaking time: overnight
Preparation time: 10 minutes
Cooking time: 1 hour
Serves: 8

Ingredients:
2 medium onions
2 cloves garlic
1 cup chickpeas, soaked overnight
1 cup white beans (such as cannellini or haricot), soaked overnight
1½ quarts chicken or vegetable stock
1 bay leaf
½ teaspoon cumin
½ teaspoon paprika
1 large bunch silverbeet
salt and pepper
7 ounces fresh ricotta
olive oil

BEANS, LENTILS, & LEGUMES

"Red Beans and Ricely Yours."

—Sign-off used by Louis Armstrong

MANY YEARS AGO, AS A YOUNG WHIPPER-snapper university student, I embraced lentils and legumes purely because of their cost. I remember walking past them in the supermarket aisle (in a dank and lonely corner of the supermarket, I might add) and noticing that a 17½-ounce packet of lentils cost less than a candy bar. It hit me that with a single packet, I could make several meals for less than the cost of a pizza.

Beans, lentils, and legumes are not only cheap; they are also extremely good for you. These foods have nourished peasant populations for millennia, and are still a staple food in many third-world countries. Throughout history, they have been considered a "poor-man's food" and have often replaced meat and fish when these ingredients were not available or too expensive. Poor Catholics who could not afford fish during Lent ate lentils instead.

To be truly delicious, most legumes need a bit of extra flavor, usually in the form of herbs, spices, garlic, or bay leaves. Traditional dishes often combine beans and legumes with some kind of protein or fat—ham hock with split-pea soup, lentil soup with black pudding, or bean soup with bacon and silverbeet. They also go beautifully with fresh herbs and produce from your garden.

There are many varieties of legumes available, wherever you travel. But for the purposes of the following pages, we'll consider the varieties most commonly found in supermarkets and organic foodstores. With a small bag of any of these, you can make cost-effective and nourishing soups, stews, and salads. There is a wonderful variety of dishes to be made with each of these legumes, so let's get cracking on soaking, sprouting, souping, salading, blanching—and, of course, eating …

Preparing your legumes

Most beans and legumes require a period of pre-soaking before they are cooked, as they contain large amounts of carbohydrates called oligosaccharides, which human digestive enzymes can't readily convert into absorbable sugars. Pre-soaking causes these sugars to be broken down and released into the soaking water. Lentils and split peas do not require pre-soaking, but all of the larger beans—kidney, chickpea, adzuki, and white beans—do.

Pre-soaking beans is easy—place the beans in a bowl of cold water and leave them there for twelve to twenty-four hours. The longer the pre-soaking time, the shorter the cooking time, and the more tender the bean is once it is cooked. If you pre-soak for more than twenty-four hours, it's a good idea to change the water once or twice. To speed up the pre-soaking process, you can use hot or boiling water and/or make the pre-soaking water alkaline by adding a small quantity (about a teaspoon) of bicarbonate of soda for every quart of water. Make sure that you drain and rinse your beans well before you cook them (the soaking water must be discarded). You can keep the cooking water to use as a vegetable stock in other dishes. If legumes are not soaked properly, you may notice a bit of bloating or gas following your meal (yes, that's right. Also known as a fart).

Of course, for any of these recipes, canned beans can replace dried and pre-soaked beans. Just keep in mind that canned beans are not nearly as cost effective and for some reason, just don't taste as good.

A NOTE ON SOYBEANS

Soybeans are an anomaly in the world of legumes. These beans were traditionally fermented for a period of months, sometimes years, to make old-style condiments such as miso, natto, and soy sauce. Asian cultures knew that the beans required fermentation to naturally break down components that were indigestible and caused ill health effects. Not surprisingly, soybeans have been found to be rich in phytates, which work as anti-nutrients (these attach themselves to nutrients and draw them from your body). They also contain large amounts of protease inhibitors and oligosaccharides, which interfere with digestion. So try to enjoy soybeans as they were traditionally consumed—that is, fermented to produce such products as miso, soy sauce, natto, and tempeh.

cannellini bean salad with pumpkin & beets

I love the combination of baked beetroot and pumpkin, and this is a terrifically easy salad to put together. With a few beets from your garden and some preserved artichokes and dried beans from your pantry, you are ready to go.

Pre-soak the beans as described earlier in this chapter.

When you're ready to start cooking, preheat your oven to 350°F. Grease a baking tray with a little cooking fat.

Rinse the beans and place them in a medium saucepan. Cover them with water and bring to a boil. Add the bay leaf and garlic, then reduce to a gentle simmer and cook, partially covered, for 45 minutes or until the beans are soft.

Meanwhile, scrub the beetroots and remove the tops, tails, and any gritty parts from the skin. Chop the beetroot and pumpkin into bite-sized chunks and place them on a baking tray. Sprinkle with salt, then bake for 20 to 30 minutes or until the pumpkin is soft and lightly browned. Remove the tray from the oven and leave to cool.

When the cannellini beans are soft, remove them from the heat and drain the cooking water. Put the beans in a large

Preparation time: 10 minutes
Cooking time: 45 minutes
Serves: 6

Ingredients:
1 cup cannellini beans, soaked
   overnight
1 bay leaf (optional)
1 clove garlic (optional)
3 artichoke hearts preserved in
   olive oil or brine
½ pound fresh green beans
   (about 2 cups)
1 medium beetroot
1⅓ pound Japla pumpkin
1 handful fresh basil
salt and pepper
1 teaspoon white-wine vinegar
1½ tablespoons olive oil

salad bowl. Sprinkle with sea salt and a dash of olive oil, toss well, and leave to cool.

While the other ingredients cool, top and tail your green beans. Use a vegetable steamer to steam them for 5 minutes or until they are lightly cooked. Rinse them in cold water and cut them into thirds.

The salad can be served either warm or chilled. When everything has cooled to the desired temperature, combine the green beans, beetroot, pumpkin, and cannellini beans. Finely slice the artichoke hearts and stir them through, along with the chopped basil, vinegar, and olive oil. Season to taste and serve.

TREACLE BAKED BEANS

These are cheap as chips to make, and always a winner on a cold winter's night. Serve them with a generous leafy salad and some crusty bread, and you have a delicious meal-in-one.

Soak your beans overnight, as described earlier in this chapter.

After soaking, rinse the beans and place them in a large soup pot. Finely chop the onion and add it to the pot along with the stock, bay leaf, and 2 cups of water. Turn the heat up and gently simmer for 1 hour or until the beans are tender. They will double in size, so you'll need to have plenty of liquid in the pot for them to soak up; make sure all the ingredients are completely immersed, and check on the pot occasionally as it cooks.

Meanwhile, prepare the allspice and cloves by pounding them in a mortar and pestle to form fine crumbs.

Add the spices to the dish, along with the garlic, treacle, Worcestershire sauce, soy sauce, and tomatoes. Continue to simmer for another hour and 10 minutes. The beans should be tender and the liquid should boil down to form a thick and delicious paste.

Season with salt and pepper to taste, then serve with a generous handful of parsley and a dash of olive oil.

Soaking time: overnight
Preparation time: 10 minutes
Cooking time: 2¼ hours
Serves: 8

Ingredients:
1½ cups (17½ ounces) white beans, such as haricot or cannellini, soaked overnight
1 large onion
2 cups stock
1 bay leaf
2 cloves garlic, crushed
½ teaspoon whole cloves
½ teaspoon whole allspice berries
1 tablespoon treacle
2 teaspoons Worcestershire sauce
6 large juicy tomatoes, cut into quarters, or 1 can (16 ounces) whole tomatoes
2 teaspoons soy sauce
1 handful fresh flat-leaf parsley, finely chopped
olive oil

CHICKPEA SALAD WITH GREENS

This is one of my favorite take-it-to-work lunches. Even after bumping along in my lunchbox on the back of my bike, it still tastes wonderful.

Pre-soak the beans as described earlier in this chapter.

When you are ready to start cooking, drain and rinse the chickpeas and place them in a small saucepan. Cover them with water and place the pot over medium heat. Gently simmer for 50 to 60 minutes, or until the chickpeas are soft.

Meanwhile, break the broccoli into large chunks and slice the cavolo nero into bite-sized pieces. In a small vegetable steamer, cook them for 5 minutes or until they are just tender. Rinse them under cold water and pat them dry, then cut the broccoli into small florets.

When the chickpeas are ready, drain them from their cooking water and place them in a salad bowl. Add the vegetables and cheese, season to taste, and toss through the vinegar and olive oil. This salad can be served warm or cold.

Soaking time: overnight
Preparation time: 10 minutes
Cooking time: 1 hour
Serves: 5

Ingredients:
1 cup chickpeas, soaked overnight
1 large broccoli
1 cup cavolo nero
3½ ounces goat's cheese or soft feta cheese
2 teaspoons apple-cider or white-wine vinegar
¾ tablespoon olive oil

Lentil Salad

Alone, this is a satisfying light lunch or mid-morning snack. It's also a delicious accompaniment to a main meal.

Rinse the lentils, then place them in a medium saucepan with the bay leaf and garlic. Cover them with water and bring to a boil, then reduce the heat and simmer for 45 minutes, or until the lentils are soft.

Drain the lentils and put them in a salad bowl. Add the vinegar and olive oil. Toss well and season to taste. Put the bowl in the refrigerator and leave the lentils to cool.

Finely dice the capsicum and celery and finely chop the fresh herbs. When the lentils have cooled, add the herbs and vegetables and toss well. Crumble the feta and stir it through. Serve at room temperature.

Preparation time: 10 minutes
Cooking time: 1½ hours
Serves: 8

Ingredients:
2 cups French green lentils
1 bay leaf
1 clove garlic
2 teaspoons white-wine or apple-cider vinegar
¾ tablespoon olive oil
1 red capsicum
2 large celery sticks, including some of the leaves
1 small handful mint, finely chopped
1 small handful basil, finely chopped
5⅓ ounces feta cheese

SPLIT-Pea Purée (Fava)

The ancient Greeks used to purée split peas into a dip much like this one here. Accompanied by a green leafy salad and an array of meat and fish dishes, it made for a wholesome meal. Split peas do not require any pre-soaking, so this is an easy dish to whip up at the last minute.

Put the split peas, garlic, paprika, and cayenne pepper in a saucepan. Cover them with water and bring to a boil. Reduce the heat and simmer for 50 minutes, stirring occasionally and adding extra water if necessary. The peas should turn to mush and develop a nice gooey texture.

When the peas have cooked, allow them to cool to room temperature, then stir through the lemon juice and olive oil. Season to taste and drizzle with a little extra olive oil to serve.

Preparation time: 10 minutes
Cooking time: 50 minutes
Serves: 8

Ingredients:
2 cups yellow split peas
2 cloves garlic
½ teaspoon paprika
¼ teaspoon cayenne pepper
1 large lemon
¼ cup olive oil
salt and pepper

cannellini bean dip

Dips are for more than dipping—they can be used on sandwiches, or served on toast with avocado or ham. I love making up a large batch when we are having people over for dinner. The leftovers can be relished for days, livening up work lunches or picked at with bread or crackers.

After pre-soaking the beans, rinse them, then put them in a large pot, and cover them with water. Add the bay leaf and garlic. Bring to a boil, then reduce the heat and simmer for 40 minutes or until the beans are soft.

Strain the beans from their cooking water. In a food processor, combine them with the juice of the lemon, a thin curl of lemon rind (about an inch long), the anchovies, parsley, and salt and pepper to taste. Gradually add the olive oil in a thin and steady stream, while pulsing the food processor on a high setting. The beans should clump together to form a smooth paste. Add a little extra olive oil if you need to.

Season to taste, then transfer the dip to a bowl or airtight container and refrigerate. The dip will keep for a good couple of days in the fridge.

Soaking time: overnight
Preparation time: 10 minutes
Cooking time: 45 minutes
Makes: 1 generous bowl

Ingredients:
- 1 cup cannellini beans, soaked overnight
- 1 bay leaf (optional)
- 1 garlic clove (optional)
- 1 medium lemon
- 3 anchovy fillets in extra virgin olive oil, or 1 preserved sardine
- ½ cup fresh chopped parsley
- salt and pepper
- ½ cup olive oil, and perhaps a little extra

SWEET POTATO HUMMUS

I love this dip. I like to make up a big batch and use it throughout the week in sandwiches and as a snack with sliced carrots and celery. It's easy and economical, and can keep me going through a long week.

If you've soaked your own chickpeas, drain them, and put them into a medium saucepan. Cover them with water, bring to a boil, then simmer for 30 minutes. If you're using canned chickpeas, you can skip this step.

Coarsely chop the carrot and sweet potato and add them to the pot along with the garlic. Simmer for a further 20 to 30 minutes, or until the chickpeas and vegetables are soft.

Drain the chickpeas and vegetables and, using a food processor, combine the chickpeas and vegetables with the spices, lemon juice, and salt to taste. Add the olive oil gradually, puréeing the mixture at a high speed until a smooth paste forms. This quantity of olive oil is only approximate—you may need a little more or less to get the desired consistency.

Season to taste and serve, or refrigerate in an airtight container.

Soaking time: overnight
Preparation time: 5 minutes
Cooking time: 1 hour
Serves: 8

Ingredients:
1 cup dried chickpeas, or 2 cans (16 ounces each) chickpeas
1 large sweet potato
1 medium carrot
1 clove garlic (or more if you particularly like garlic)
½ teaspoon paprika or Ras el Hanout spice mix
juice of 1 large lemon
sea salt
½ cup olive oil

MEAT

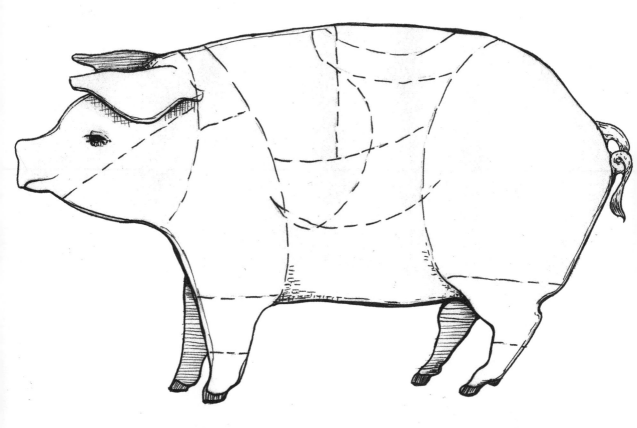

"The only thing wasted on a pig is his squeal."

—Mrs. Burns, dairy farmer, Gippsland, Australia.
She died aged ninety-four while still running
her own farm and tending her backyard tomatoes.

HOW DO YOU FEEL ABOUT THE MEAT YOU buy? Are you happy to buy something that is anonymously labeled, packed in plastic, and stacked under lights at your local supermarket? Or would you prefer to know a bit about your animal—where it comes from, how it lived, and the journey it took to get to your dinner plate?

As consumers, we have a great power to influence how our meat is produced, packaged, and distributed. We can demand better quality meat by connecting to our local farms, supporting good butchers, and telling retailers what we would like to see on their shelves. Purchasing from local, small-scale farms is the best way to support your local food economy. If you shop at farmers' markets or buy directly from the farm, try to buy your meat in bulk and look for unusual cuts and different breeds. For instance, why not try some mutton instead of lamb, or look for boiling fowl instead of the typical plumped-up hen next time you shop?

Fresh from the farm

When you buy your meat straight from the farm, you can see directly where your food is coming from. You can also get a better price, have access to a range of different cuts, and be directly involved in your local food economy. For me, the knowledge that an animal has lived well is far more important than any organic or biodynamic certification tick.

By shopping directly from the farm you can also save money, as you cut out the transportation and retail costs. Buying from the farmer is generally 50 percent cheaper than shopping at a butcher's shop or organic store. The farmer is also getting a better price; for most farmers, it is much more profitable to sell directly to consumers than through retail outlets.

I also enjoy seeing the diversity that is available at the farm gate whenever I visit—a fresh rooster, older hens, mutton, fresh spring lamb. Before I started dealing directly with farmers, I never realized how many ways there were to cook meat, using different breeds or animals of different ages and different cuts. What a shame that more of this diversity is not available in the mainstream retail market!

Shopping directly from the farm often requires that you buy your meat in bulk. You can store any excess in your freezer and explore cooking with the whole animal, including all the "thrifty bits."

Butchers

Good butchering is a highly specialized artisanal craft. Traditionally, it took years of training and butchers were respected pillars of small communities. A good butcher will have a clear understanding of where his meat came from and will cut, carve, and age his meat on-site at the back of his shop.

Unfortunately, there aren't many good butchers left. They've been replaced by "meat carvers," who lack the intimate knowledge required to choose the best quality meat and prepare it to its full potential. The blame for this lies with consumers, who increasingly seem to want only certain "easy" cuts of meat—racks of lamb, loin chops, beef steaks, and chicken breasts. We've forgotten how to cook many of the more economical cuts—animal feet, bones, offal, fat, and stewing steaks. Fifty years ago, housewives sought out these cuts and nothing went to waste. They were also fussy about the quality of what they purchased; they noticed if a slice of liver wasn't fresh or a stewing steak lacked gristle. If the quality was lacking, the butcher wouldn't get their business the following week.

Nowadays, with consumers buying only a few of the leaner cuts, many butchers get their meat ready-cut from the slaughterhouse. The process of dry-aging (where the meat is hung out to mature behind the shop, giving it a richer flavor) has been abandoned by many butchers, replaced by "cryo-vaccing" to save time and money. At the same time, stricter regulations mean butchers are no longer per-mitted to make their own salami or air-dry their own hams in the traditional way.

My interest in finding a good butcher started when I embarked on a quest to cook with traditional Jewish-style sweetbreads—thymus and pancreas glands—but couldn't find them at my local store. I tried everything. I attempted an "offal drive" with several like-minded friends; we all pitched in and tried to buy them in bulk from a local slaughterhouse. No go. I tried a local farmer, but he couldn't sell them to me either. Finally, I found a good butcher. And when I say "good," I mean really good. Roger processes whole carcasses in the back room of his shop, so I can access all the different cuts, including other hard-to-find items like suet, chicken feet, and pork fat. He is a godsend.

These were all once kitchen essentials, but they've been utterly lost from our food supply. As consumers, we've become unadventurous. Consequently, there is now a much narrower supply of meat on the market. For this to change, we need to be prepared to buy more of our food in bulk, and to source our meat directly from the farm or from an excellent butcher. A good butcher is one who knows how to prepare an animal from top to toe. Nothing is wasted at a good butcher's shop, and you should be able to buy all the cuts—not just the popular ones. A good butcher may have his own farm, or at the very least have a close relationship with wherever his meat comes from. He will be able to tell you clearly about the origins of your meat, how it was raised, and how it was processed—was it wet or dry-aged, and for how long? He will do most of his butchering and processing onsite. If he dry-ages his own meat and makes his own salami and hams, he is a true gem. So when you find one, don't let him go!

SUPERMARKETS

Supermarkets are not known for careful processing, or for sourcing the best produce. That's a given. But I have noticed that some smaller supermarkets, especially locally run co-ops, provide detailed information about the quality and history of their meat via information sheets on display next to the produce. This is not as good as a proper chat with your butcher or farmer, but I would say it is the next best thing.

On a larger commercial scale, supermarkets are supplying more of the frugal cuts—such as lambs' necks, chicken carcasses, pigs' feet, and stewing steak—in response to consumer demand. They are also stocking a more diverse range of produce from organic and grass-fed farms. This is good. But most of the meat they sell is still anonymously labeled, with no details about where it came from. It always makes me want to ask: What was the farm like? What did the animal eat? Did the farmer get a good deal? If you are shopping at supermarkets, look for thrifty cuts, organic or grass-fed, and try to support local farms wherever you can.

WHAT TO BUY?

Many people are confused as to what constitutes good meat. Should you buy grain-fed or grass-fed? Free-range or organic? From a nutritional perspective, the healthiest animal products come from animals that have lived as close to their natural environment as possible. For livestock, this means a diet of grass and hay, preferably involving "salad-bar grazing" on a diverse range of grass varieties. For poultry, the ideal diet includes an omnivorous range of grasses, insects, and grubs for protein, and only minimal grains.

Nutritional tests have shown that animals raised on natural diets have fewer medical problems and provide better-quality produce. Their products contain more omega-3 and conjugated linoleic (CLA) fatty acids, both healthy fats known to benefit heart health and aid in the prevention of inflammatory and autoimmune disorders. Grass-fed and free-range products also contain higher levels of antioxidants and important fat-soluble vitamins such as A, D, and E. Grass-fed livestock do not suffer from many of the afflictions faced by their grain-fed counterparts such as acidosis, rumenitis, liver abscesses, and bloat, and are at a reduced risk of *E. coli*

contamination. And when the animals are healthier, they require fewer antibiotics, again resulting in healthier produce.

So if you are looking for the healthiest meat, look first at what the animals have eaten. Check that they had access to appropriate food, and also that they enjoyed fresh air and exercise. Foods that are not appropriate for animal consumption include large amounts of corn and grains (for livestock), or any soy meal, brewery waste, and what the food industry likes to call "miscellaneous food scraps."

CERTIFIED PRODUCE

Certification systems are really a secondary measure, used to ensure quality when we cannot see for ourselves how the animal has lived. Personally, I would much prefer to buy fresh produce from a farmer whom I know, rather than an anonymous package with an organic stamp.

But if you are buying meat from a retail outlet, the first thing you should try to determine is whether the animal was raised using natural feeding practices (i.e., look for "grass-fed," "free-range," or "pasture-raised"). If this is well established, then you can look for organic or biodynamic certification. Organic and biodynamic farming methods don't rely on chemicals or pesticides. Instead, nutrient-dense grasses and feed are used both to nourish the animals and enliven the soil.

The bottom line is, find out where your food comes from. Certification systems are a great way to do this, but not the only way, and often not the cheapest way. If you can connect with your local supplier and see that their animals have enjoyed a natural diet and a decent dose of air and sunshine, don't worry too much about whether or not they are officially "certified"—just use your own judgment and common sense.

Buying in bulk

If you have a good freezer, purchase a whole or half animal, or buy in bulk. This applies to all types of meat: it might involve buying a whole chicken or duck, rather than just the drumsticks or breasts, or purchasing half a sheep carcass and having it butchered. This is cost-effective and less wasteful, and gives you access to a wider range of cuts. Buying the whole animal also gives you access to all the wonderfully nutritious "extra bits." I am not a vegetarian (obviously!). But I do believe that if we are going to eat meat, we should at least respect the life of the animal and what it has given to us. Wasting no part of it is an important aspect of this.

To buy in bulk directly from the farm, you can negotiate an arrangement at local farmers' markets, or tap into local grassroots movements such as buying clubs and co-ops. I have an arrangement with a local farmer whereby I purchase a whole animal (if it's a larger animal, I usually go in with a few friends). We usually go home with a few coolers full of meat, enough to last us for a few months.

When buying half a cow (a top-of-the-range, grass-fed organic beast), we usually pay around half of what we would at a local retail outlet. Prices vary, so you'll have to do some research and see what your local farmers have to offer. You should also check your local regulations. Generally speaking, if you can get a number of people together and buy a large quantity in one go, it will be less trouble for the farmer, meaning a better price for you.

If you're shopping at supermarkets or organic retailers, look for a whole chicken rather than just the lean breasts or drumsticks. The price is often comparable, and you'll get three meals rather than one.

Thrifty bits

A generation or two ago, it was standard to prepare a casserole or some other slow-cooked dish a couple of times a week. This could use up all those odd cuts of meat—the stewing steak, oyster blade, lambs' necks, or oxtail, or whatever happened to be cheap that week at the butcher's. Nowadays, with so many of us working longer hours, we have come to favor cuts like stir-fry, chops, and more expensive steaks, which don't need so much cooking time. The drawback is that these cuts are a lot more expensive, and much less rich in gelatin and nutrients. To get the best meat for the lowest price, look for the less popular pieces— the stewing steaks, offal, and odd bits (tails, bones, feet, etc.) and make the most of them. These cuts are not only delicious; they are also extremely good for you. Not surprisingly, these odd bits have always been favored in traditional cuisines. Pigs' feet are a European delicacy. Rooster combs are *haute cuisine* in southern France. And heads, feet, and wings provide gelatin, which is essential to stocks and stews.

Stewing steak (also called oyster-blade, chuck steak, and casserole steak): These cuts come from the muscular part of the animal such as the legs and neck. They are thick with muscle and rich in gelatin, which means they require long, slow, gentle cooking. They are highly nutritious and are delicious in casseroles, stews, and other slowly cooked dishes.

Mince: Mince is very cost-effective, and many mince recipes are extremely quick and easy: think hamburgers, meatballs, and meatloafs. If you are trying to save on grocery bills, have a few good mince recipes up your sleeve and buy it in bulk.

Offal and "odd pieces" (feet, tails, necks, and so on): Offal is significantly cheaper than the leaner cuts such as steak or shoulder roast. These cheaper cuts were considered the most nutritious parts of the animal by many traditional groups. Offal has the highest amount of fat-soluble vitamins (vitamins A, D, and K), while cuts such as liver are rich in iron and other important vitamins (pregnant mothers are still encouraged by nutritionists to eat liver pâté to keep their iron levels up). "Odd pieces" such as the tail are also rich in gelatin and important minerals, and are great in soups and stews. See "Cooking with Offal" later in this chapter for detailed information about different cuts.

Bones: Animals have a high ratio of bones to meat, so throwing the bones away means much of the animal goes to waste. Because it is rich in gelatin, bone stock is extremely nutritious. You can buy bones cheaply from your butcher, or in bulk from your local farm. See the "Soup" chapter of this book for a multitude of ways to prepare stocks and stews. If you make stock in advance, you can freeze it and use it to make cheap and delicious meals throughout the week.

COOKING WITH OFFAL

Sweetbreads: These include the pancreas and thymus glands. To prepare, soak them in cold, salted water for two or three hours. Drain them, then place them in boiling water with a bay leaf, some peppercorns, and ¾ tablespoon of vinegar. Simmer for 15 minutes. Allow to cool, then remove any traces of skin or sinew.

The sweetbreads can now be cooked in a creamy sauce, sautéed with mushrooms, or fried with some fresh herbs.

Brains: These need to be soaked in salted water for at least 3 or 4 hours. After soaking, drain away the liquid and remove any membrane or specks of blood. Place the brains in boiling water with a bay leaf, some peppercorns, and ¾ tablespoon of vinegar. Cook for 20 minutes on a gentle simmer. Like the sweetbreads, they can now be sauced with mushrooms, butter, or cream. A gentle fry-up with some butter and fresh herbs is also a treat.

Liver: The key to cooking liver is to make sure that it's fresh. Then try to slice it as thinly as possible—I often get my butcher to do this for me. I like to soak liver in some cold milk for 10 to 20 minutes before tossing it in flour and frying it with a little cooking fat on high heat. Fry for only a few seconds on each side; it should be only lightly cooked.

Kidneys: Lamb or calf kidneys are lighter tasting and only require gentle cooking. I've enjoyed these barbequed or lightly fried with some fresh herbs from the garden and a salad. Ox kidneys or the kidneys of mature cattle tend to be tougher, so do better with long, slow cooking (for instance in a steak and kidney pie) or sautéed with a rich red-wine sauce.

Heart: Trim away the veins and arteries so that you are left with just the muscle itself. Leave the fat on, as this helps to keep the meat tender as it cooks. I recommend cooking this just as you would a casserole, with plenty of herbs, stock, and wine. Gentle cooking at 350°F for 3 to 4 hours should do the trick.

Tongue: The tongue is also a muscle and needs to be brined for at least 24 hours, then gently cooked with bay leaves, peppercorns, vinegar, and herbs for 3 to 4 hours. It is wonderful served cold with a tangy sauce.

beef casserole

This recipe comforts and warms, and turns a few simple ingredients into a hearty and wholesome meal. During the week, I often halve this recipe and use a smaller cooking pot to make a quick and easy dinner for two. It goes well with baked spuds and a fresh green salad from the garden.

Preheat the oven to 350°F. Place an ovenproof pot with a tight-fitting lid (cast-iron is best) on the stovetop over medium heat. Add the wine, stock, and tomatoes and bring to a gentle simmer.

Combine the arrowroot and cumin in a small bowl and toss the meat in this mixture.

Place a frying pan over low heat and gently cook the onions in a little fat until they soften and brown (between 5 and 10 minutes). Add these to the cooking pot and scrape the frying pan clean.

Return the pan to the stove and turn up the heat. Add a teaspoon of fat and then fry the meat until it browns, sealing in the flavor.

Preparation time: 5 minutes
Cooking time: 2½ hours
Serves: 4 (generously)

Ingredients:
½ cup red wine
2½ cups stock
6 large ripe tomatoes, coarsely
    chopped
2 tablespoons arrowroot
    powder
1 heaped teaspoon dried cumin
1¾ pounds stewing steak, cut
    into bite-sized pieces
fat for frying
1 brown onion, finely chopped

Transfer the meat to the cooking pot and ensure that the liquid covers the meat completely; add a little extra water if you need to, but keep in mind that the tomatoes will release their liquid as they cook. Season to taste.

Cook the casserole for 2½ hours or until the meat is tender when poked with a fork. Serve with baked potatoes and a green leafy salad.

Irish stew

For less than the cost of a delivered pizza, this dish can easily feed a family of four and provide leftovers the next day. Irish stew was traditionally made with lamb or mutton and the cheapest, most readily available cuts such as neckbones, shanks, and other offcuts. Arrowroot, cumin, and lemon juice are not traditional Irish ingredients; I've added them to thicken the sauce and provide a lighter, sweeter flavor. I hope Irish readers will forgive me!

Preheat your oven to 350°F.

Toss the meat in the arrowroot powder and cumin. Put a frying pan over high heat and add a small dollop of cooking fat, then fry the meat on both sides until it browns.

Transfer the meat to an ovenproof pot with a tight-fitting lid (cast-iron is best). Add the barley, onions, garlic, and stock and plenty of water to cover them. Cook in the preheated oven for 2 hours.

Preparation time: 15 minutes
Cooking time: 3 hours
Serves: 5 (generously)

Ingredients:
1 lamb neck, cut into chops (roughly ¾ pound)
2 teaspoons arrowroot powder
1 teaspoon cumin
fat for frying
¼ cup pearled barley
2 large onions
2 cloves garlic
1 cup beef stock
1 turnip
2 medium carrots
2 large potatoes
juice of 1 lemon
1 heaped cup finely chopped flat-leaf parsley

Meanwhile, prepare the turnip, carrots, and potatoes by giving them a good scrub and chopping them into bite-sized pieces. (You can peel them if you prefer, but keep in mind that a lot of the nutrients are in the skin.) Add them to the pot and check that there is enough liquid to cover them.

Return the pot to the oven and cook the stew for a further hour. It's ready when the the meat is tender (give it an extra 10 minutes if it doesn't seem quite soft enough).

Once the stew is out of the oven, loosely pick the meat from the bones and stir this through the stew. Season generously with salt and pepper and stir through the lemon juice and flat-leaf parsley. Serve immediately.

POT-au-Feu

Pot-au-feu translates literally as "pot on the fire." This recipe has a long history in French cuisine. Traditionally, poorer families would have had to make do with barley or rice, a few root vegetables, plenty of bones, and probably very little meat. Nowadays, this dish is still a cost-effective way to make the meat you buy go a little bit further, producing not just one but three courses. With long, slow cooking, even the toughest cuts become beautifully tender. Despite its humble origins, pot-au-feu is still a favorite in France and served at many top-notch restaurants throughout the countryside. Serve it with crusty bread and some mayonnaise, aioli, mustard, or horseradish sauce.

Place a large stockpot on the stove. Add the meat, excluding the marrow bones, and cover with water. If you are using osso bucco, cut the bone out from the steak and set it aside, and add just the meat to the pot. If you are adding any additional cooking bones, add them now. Cover them with water.

Bring the water to a boil, then reduce the heat and simmer very gently, partially covered. As it cooks, a layer of scum may form at the top; you can skim this off with a wooden spoon and discard it. Continue gently simmering for 3 to 4 hours so that the meat can soften.

Preparation time: 5 minutes
Cooking time: 6 to 7 hours
Serves: 6

Ingredients:
- 5½ pounds bony cuts of beef (I recommend 2¼ pounds beef brisket, and 2¼ pounds marrow bones. But you can use any cuts you like, including tongue, ox cheek, or oxtail). If you have room in your pot, add a few extra bones for additional flavor
- 1 bouquet garni (or 1 teaspoon dried thyme, rosemary, and basil)
- 2 bay leaves, fresh or dried
- ½ cup brown rice or barley (optional)
- 2 large tomatoes
- 4 large carrots, topped and tailed but left whole

2 large leeks, cut into 2-inch
 pieces
2 medium onions, cut in half
2 turnips, cut in half (optional)
1 bulb garlic
1 large bunch flat-leaf parsley

Add the herbs, rice, and vegetables (excluding the parsley and garlic) and continue to cook gently for an additional 1½ hours.

Add the marrow bones and cook for 20 minutes or until the fat in their centers is soft and runny.

To serve your first course, remove the marrow bones from the pot and place them on a serving dish with some crusty bread, fresh garlic cloves, and a handful of fresh parsley. Encourage your guests to rub the bread with garlic, scoop out the marrow, spread it over the bread, and sprinkle with salt and parsley.

For your next course, drain the broth from the meat and vegetables. Serve the broth with a little rice and a few vegetables in each bowl. Season to taste.

For the main course, arrange the meat and vegetables on separate platters and serve with mayonnaise, aioli, mustard, and some more crusty bread.

This dish can also be prepared in advance—the meat and vegetables served cold, and the broth and marrow bones re-heated before serving. This allows you to skim off the fat that rises to the surface when the broth is cooled.

home-cured bacon

Home-curing bacon is easy and extremely cheap. Pork belly is a terrifically thrifty cut, and home-curing it will preserve it for a period of months. Snippets of bacon are delicious in soups and stews, or fried with some eggs for breakfast. You can also boil it to make a traditional French dish known as petit salé. All the quantities in this recipe are approximate. The idea is to use enough salt and sugar to draw the water out from the meat, so go ahead and add a bit more if it looks like it needs it.

Use a sharp knife to cut some slits, about ¼ or ½ inch deep, in the pork meat. Combine the salt, bay leaves, thyme, sugar, and peppercorns in a large bowl and toss well. Rub this mixture all over the meat.

Layer the pork belly one piece on top of the other in the bowl. The meat should be completely covered by the salt, so spread the salt around and add a little more if you need to. Place a sheet of greaseproof paper over the top and weigh it down with something (I usually use a bag of dried beans or oatmeal). Place the bowl in the refrigerator and leave it for 48 hours.

Preparation time: 10 minutes
Resting time: 5 to 7 days

Ingredients:
½ of 1 large pork belly, cut into thirds
17½ ounces fine sea salt
3 or 4 bay leaves, finely chopped
1 large handful fresh thyme, finely chopped
½ cup brown or whole cane sugar
2 tablespoons black peppercorns, roughly crushed in a mortar and pestle

Remove the bowl from the fridge and drain off any excess water. Rearrange the meat so that the bottom slice is on top and the top slice is on the bottom. Pack any excess salt back onto the meat (if the salt doesn't seem enough to cover the meat, make up another mixture of salt and sugar). Cover the meat with the paper again, return the bowl to the fridge, and leave it for another 48 hours.

After 5 days, your meat should be ready to nibble on. I like to cure mine for 6 or 7 days, checking on it every 2 or 3 days. The longer it is cured, the tougher and saltier it will become and the longer it will last in your fridge or pantry.

When you have finished the salting process, you can wrap the bacon in a natural fiber (such as cotton or muslin), or fold some greaseproof paper over it and store it in a cool place such as a fridge or cellar. It should keep for a good couple of months. Slice off thin slices whenever you need it. If you have salted it for a long period of time, you may have to soak it in fresh water for an hour or two before using it to soften it up.

Home-cured Bacon served with Lentils

This is a staple dish in northern France, where it is known as petit salé aux lentils. It is an excellent and thrifty way to use up home-cured pork belly. If your bacon seems too tough or salty, soak it in filtered water for a few hours before cooking this dish.

Place the pork belly, vegetables, and herbs in a saucepan and cover them with cold water. Bring to a gentle simmer and cook for 2 hours or until the meat is tender and soft. Remove the pork belly from the pot and set aside to cool. Discard the vegetables but retain the cooking liquid.

Add the lentils to the pot with the cooking liquid and turn up the heat. Simmer for 40 minutes or until tender.

Drain the lentils, season them with salt and pepper to taste, then transfer them to a serving dish and place the pork belly on top.

Preparation time: 10 minutes
Cooking time: 2 to 3 hours
Serves: 2

Ingredients:
1⅓ pounds home-cured pork
  belly
2 or 3 carrots
2 or 3 celery sticks
1 onion
1 bouquet garni (or ½ teaspoon
  each of dried thyme and
  rosemary, plus 1 bay leaf)
1 cup French green lentils

MUTTON CURRY

Mutton (the meat from older sheep) has fallen out of favor in recent times, as sweeter and more expensive cuts of lamb became more widely available. This is a shame, as well-prepared mutton has a delectably subtle flavor. It requires longer, gentler cooking than lamb and is best combined with plentiful spices and herbs. If you can't find mutton chops, you can substitute lamb loin chops, which are one of the cheapest cuts of lamb.

Preheat the oven to 350°F.

Place a frying pan over high heat and fry the chops with a little cooking fat until they brown, sealing in the flavor. Transfer the meat to a medium-sized casserole dish and coat it with the curry powder.

In the same frying pan, fry the onions over low heat for 5 to 10 minutes, or until they caramelize. Add them to the pot with the meat, along with the coconut milk and bouquet garni. Stir well.

Place the cooking pot in the preheated oven and cook at 350°F for 1¼ hours. Remove from the oven and stir through the yogurt. Season to taste and serve with brown rice and a fresh leafy salad.

Preparation time: 5 minutes
Cooking time: 1¼ hours
Serves: 4

Ingredients:
1¾ pounds mutton or lamb
    chops
fat for frying
4 teaspoons curry powder
1 medium onion, finely
    chopped
1¼ cups coconut milk
1 bouquet garni (or ½ teaspoon
    each of dried rosemary,
    thyme, and marjoram)
⅓ cup plain yogurt

OXTAIL STEW WITH APPLES & SPICES

This hearty stew might take a little longer to prepare than your average dish, but don't be put off; it's delicious and well worth the effort. Oxtail is a traditional cut that seems to be coming back into fashion. Even at my neighborhood middle-of-the-road supermarket, I've spotted locally sourced oxtail on sale for less than the cost of 2 pounds of mince. It requires slow, gentle cooking for the best results. I highly recommend cooking this the day before you eat it, so that you can skim off the fat and get a rich and gelatinous sauce. The vegetables don't need to be super fresh for this dish, so it's a good way to use up anything that's starting to look a little wilted. Serve with rice and a leafy green salad.

Preheat your oven to 300°F. Put a frying pan over a high heat, add a little cooking fat, and fry the meat in batches to seal in the flavor.

Place the stock, wine, and bay leaves in a heavy-based ceramic dish with a tight-fitting lid. When the meat is ready, transfer it to this pot and add enough water to cover it entirely.

Cook the stew in the preheated oven for about 5 hours, or until the meat is tender and beginning to fall off the bone.

Preparation time: 20 minutes
Cooking time: at least 6 hours
Serves: 4
Best prepared the day before

Ingredients:
2½ pounds oxtail, cut into
    1-inch pieces
fat for frying
2 cups stock
½ cup red wine
2 bay leaves
1 large onion
2 large carrots
2 large parsnips
1½ tablespoons tomato
    ketchup
1½ teaspoons Ras el Hanout
    spice mix
3 medium cooking apples
salt and pepper

Remove the pot from the oven and allow it to cool. Drain the meat from the liquid and place the liquid in a jug in the fridge. After several hours the jug of liquid will cool and a layer of fat will form at the top. Scrape this off and discard it. Similarly, any excess fat on the oxtail meat should be trimmed off and discarded.

When you're ready to prepare your dinner, put the casserole dish on the stovetop and add the meat and cooking liquid and bring it to a gentle simmer.

Finely chop the onion and fry it in a frying pan on medium heat until it caramelizes and turns golden.

Chop the carrots and parsnip into bite-sized pieces. Add these to the cooking pot with the onion, ketchup, and spice mix. Place the pot in the oven, or allow to simmer on the stove top, and cook for an additional 1¼ hours, or until the vegetables are tender.

Cut the apples into quarters and add them to the pot. Cook for a further 15 minutes (preferably on the stovetop, so that you can keep an eye on things), or until the apples are soft. If necessary, remove the lid and turn up the heat to boil down any excess liquid. Season to taste and serve.

Baked Sweetbreads with Butter & Sage

I confess I was a bit freaked out the first time I brought these little thymus glands home from my butcher. They were soft, mushy, and very weird to the touch, and I did have some reservations about cooking (let alone eating!) them that first night. But I finally drummed up the courage to dip them in egg and breadcrumbs and bake them in the oven, and this is now my *favorite* offal dish. Please don't be dissuaded by what they look like raw. Their taste is superb, and they go beautifully with a crisp green salad.

Soak the sweetbreads in cold, salted water (use roughly ¾ tablespoon of sea salt for every quart of water). If you're in a real hurry, 30 minutes will suffice, but an hour or two is better.

When you're ready to start cooking, preheat the oven to 350°F and grease a baking tray with butter.

Put the sweetbreads in a saucepan of warm water with the vinegar, peppercorns, and bay leaf. Bring to a boil and gently simmer for 15 minutes. Drain the sweetbreads and let them cool, then slice them into bite-sized pieces.

Soaking time: at least 30 minutes
Cooking time: 50 minutes
Serves: 4

Ingredients:
1 thymus gland from a calf
sea salt
butter
3 teaspoons apple-cider or wine vinegar
1 teaspoon whole black peppercorns
1 bay leaf
1 egg
salt and pepper
½ cup finely chopped sage
½ cup breadcrumbs

Combine the egg with salt and pepper to taste and whisk well. Finely chop the sage and, in a separate bowl, toss it with the breadcrumbs.

Drop the sweetbreads into the egg mixture and let them sit for 2 minutes. Dip them in the breadcrumbs, then place them on the baking tray. Bake for 30 to 40 minutes or until they are lightly browned. Be careful not to let them overcook, as they can dry out.

Serve with a green salad.

steak & kidney pie

This is a favorite at our house. If you are new to offal, this is a good place to start. You can adjust the ratio of kidney to steak if you want more or less kidney flavor. If you slice the kidneys finely enough, they may even be mistaken for mushrooms! Although it takes a while to cook, don't be put off, as this is a great leftover dish and can save you time during the week. It can be made on the weekend and reheated for weeknight dinners or packed in school lunches. The filling can also be made in advance and frozen or refrigerated until you need it.

First, prepare your oatmeal pastry.

If your butcher hasn't already done so, chop the steak and the kidney into bite-sized pieces. Fry the meat in batches over a high heat with a little cooking fat.

When the meat is sealed, transfer it to a cast-iron cooking pot. Add the wine, soy sauce, stock, bay leaves, onion, thyme, tomato paste, and mustard. Bring the liquid to a boil, then reduce it to a simmer. You may need to add an extra half cup of water to ensure that everything is submerged.

Let it simmer on the stove for 2 to 3 hours, stirring occasionally so that nothing sticks to the bottom. Alternatively,

Preparation time: 20 minutes
Cooking time: 3 hours
Serves: 6 (generously)

Ingredients:
2 quantities oatmeal pastry
   (see recipe on page 252)
⁴/₅ pound beef or lamb kidney
2¼ pounds slow-cooking beef,
   such as beef skirt or chuck
   steak
fat for frying
1 cup red wine
1½ teaspoons soy sauce
2½ cups stock
2 small bay leaves
1 onion, finely sliced
2 teaspoons finely chopped
   fresh thyme
¼ cup tomato paste
1½ teaspoons mustard
1 pound fresh mushrooms
3 to 3¾ tablespoons arrowroot
   powder
1 egg

you can place the cooking pot, with lid, in the oven and bake it at 350°F for the same period of time. I prefer to use the stove, as it lets me keep an eye on the liquid level.

While the meat is cooking, cut your mushrooms in half and pan fry them with a little fat for 10 to 15 minutes, or until they are cooked through and reduced in size.

When the beef is tender, boil down any excess fluid so that the liquid only just covers the meat and add the mushrooms.

At this point, preheat your oven to 350°C and grease a large rectangular pie dish (about 11 inches by 6 inches) with a little fat.

Remove about half a cup of liquid from the pot and combine it with the arrowroot powder. Return the resulting paste to the pot and stir well. The filling should now be thick and gluey.

Tip the filling into the greased pie dish.

Using your fingers, roll out the pastry between two sheets of grease-proof paper to make a lid. Place this on top of the pie. Whisk the egg and brush it onto the lid as a glaze.

Bake your pie for 30 minutes or until the pastry is golden.

Brined Tongue with Salsa Verde

This is a delectable dish with a tangy green sauce that really complements the saltiness of the tongue. I've served it to guests who were nervous about eating tongue and heard them raving about it years later. If you prefer, you can buy brined tongue from most good butchers. Whether you buy it ready-brined or prepare it yourself, served with salsa verde it makes for a delicious Sunday lunch.

To brine the tongue:

To make the brine, combine the water, salt, sugar, and 1 teaspoon of the peppercorns. Bring them to a boil so that the salt and sugar dissolve, then remove from the heat and refrigerate.

When the brine is no longer hot, add the tongue and leave it to soak. The longer you leave it, the stronger the salty flavor will be. If you only have 24 to 48 hours, I'd suggest using half of the brine as cooking water when it's time to cook the tongue. Alternatively, if you have more time, soak the tongue for up to 7 days and rinse it in fresh water for a few minutes before you cook it.

Soaking time: at least 24 hours
Preparation time: 5 minutes
Cooking time: 2 to 3 hours
Serves: 6 (generously)

Ingredients
FOR THE TONGUE:
1 beef tongue
1 gallon water
21 ounces sea salt
2 cups sugar
2 teaspoons black peppercorns
1 onion
2 carrots
a few sticks celery (optional)
3 bay leaves
5 cloves

Ingredients continued ...

FOR THE SALSA VERDE:
1 slice stale bread
¼ cup olive oil
1 cup parsley (either flat or
    curly)
¾ tablespoon white-wine
    vinegar
2 anchovy fillets
1 clove garlic, crushed
half a preserved artichoke
    heart
salt and pepper

After soaking, put the tongue into a large pot with the onion, carrots, celery, bay leaves, cloves, and the remaining peppercorns. Simmer for 2 to 3 hours. When it's ready, the tongue should be tender and the outer skin fairly easy to remove.

Remove the tongue from the pot and remove the skin while it is still warm (it becomes much harder when it's cold). Some cooks peel off the skin with their fingers, but I find it easier to use a small knife.

Thinly slice the tongue and arrange the slices on a plate. Sprinkle them with sea salt and place the plate in the refrigerator to cool before serving with the salsa verde.

To make the salsa, soak the bread in the oil for ten minutes, then combine all the ingredients in a food processor and blend until smooth. Season to taste.

CRISPY LIVER WITH CARAMELIZED ONION

Some people recall bad childhood experiences when I mention liver. I admit, the first time I cooked liver it was terrible—overcooked, tough, and dry. It took me years to want to try it again—but once I got it right, I became seriously hooked.

When properly cooked, liver can be exquisite. The trick is to buy it as fresh as you possibly can, preferably grass-fed and organic. If you're a first-time liver-eater, try to buy lamb's liver, as it is milder than beef or pork. If your butcher offers to slice it for you, take him up on it. Liver should be as thinly sliced as possible, and it can be difficult to do this at home if you don't have a very sharp knife and good knife skills. I like to soak the liver in milk for a few minutes before dusting it in flour and pan-frying it; this results in a subtle taste and tender texture. If you have any leftovers, you can use them in pâté.

Start by caramelizing the onions: slice them as thinly as possible and place them in a frying pan with 1 teaspoon of butter. Fry over low heat, stirring occasionally, for 10 to 15 minutes. The onion should soften, become sweet and tender, but not brown.

Preparation time: 10 minutes
Cooking time: 15 minutes
Serves: 5 (generously)

Ingredients:
butter for frying
3 onions
1 fresh lamb's liver (about 1½ pounds)
1 cup fresh milk
flour or arrowroot powder for dusting
salt and pepper

While the onion cooks, prepare the liver. If the butcher hasn't already done so, slice the liver as thinly as possible, cutting against the grain of the meat. In a bowl, soak the liver in the milk for 5 to 10 minutes.

Dust each piece of milk-soaked liver in the flour and fry for 5 seconds on each side in a small amount of butter. The inside should still be pink, but the outside brown and crispy.

Season with salt and pepper. Serve with the caramelized onion. Fruit chutney also complements this dish nicely.

LIVER PÂTÉ

Good pâté is a special treat. It can be served on thick slices of crusty bread, with crackers, or in a sandwich of fresh arugula and creamy avocado. Many people think of it as a luxury food, but in fact pâté is very quick and cost-effective to make at home.

Preheat the oven to 325°F. Line a medium-sized tureen dish with a little cooking fat.

Pulse the bread in a food processor to form fine crumbs. In a bowl, combine the crumbs with the milk and cream. Leave this mixture to soak while you prepare the other ingredients.

Finely chop the onion and garlic. Fry them with the cooking fat over low heat for 5 minutes, or until the onions are golden.

Use a food processor to combine the onion, breadcrumbs, and all the remaining ingredients, blending them until they form a smooth paste. Season to taste, then transfer the pâté to the tureen dish, and cover with a lid or a double layer of greaseproof paper.

Place the dish in a tray of boiling water and put the tray in the preheated oven. Cook for 45 to 60 minutes or until the pâté begins to come away from the edges of the dish. Refrigerate and serve cold.

Preparation time: 10 minutes
Refrigeration time: 30 minutes
Makes: 1 medium bowl

Ingredients:
1 pound lamb's liver
1 cup milk
butter for frying
5 fresh sage leaves
1 teaspoon fresh or dried thyme
½ cup cream
½ onion, roughly chopped
⅛ cup brandy
1 thin slice of bread
1 heaped teaspoon butter
½ teaspoon ground nutmeg
½ teaspoon ground ginger
salt and pepper
juice of 1 lemon

Meatloaf with red sauce

Once upon a time, just about every housewife would have had her own prized meatloaf recipe. The beauty of meatloaf is its simplicity: it uses very cost-effective ingredients and can be served in multiple ways. It's particularly good for school lunches; it's excellent in a sandwich with some fresh lettuce and arugula. This recipe is based on a traditional Jewish meatloaf with a red sauce. Some cooks bake a boiled egg in the middle; I prefer to omit the egg, but I adore the red sauce.

Preheat your oven to 350°F.

Combine the mince, onion, tomato paste, breadcrumbs, egg, and parsley in a food processor, or mix them together in a large mixing bowl. Season with salt and pepper to taste.

Grease a baking tin (preferably loaf-shaped) with a little butter. Pour in the meat mixture and bake for 45 minutes, or until the meatloaf is cooked right through.

While the loaf bakes, prepare the sauce. Put the tomatoes, sugar, and salt and pepper to taste in a small saucepan and bring to a boil. Simmer for 5 to 10 minutes, or until the mixture has boiled down to a sauce-like consistency.

To serve, slice the meatloaf and pour sauce over each slice.

Preparation time: 10 minutes
Cooking time: 45 minutes
Serves: 6

Ingredients
FOR THE LOAF:
1½ pounds minced beef
1 onion, finely chopped
1½ tablespoons tomato paste
1½ pounds breadcrumbs (or 2 small slices of bread whirred in a food processor)
1 egg, lightly whisked
1 cup finely chopped flat leaf parsley or celery leaves
salt and pepper
butter, for greasing

FOR THE SAUCE:
6 very ripe tomatoes, coarsely chopped, or 1 can of whole tomatoes
1 or 2 teaspoons sugar
salt and pepper

PORK-MINCE APPLES

There is nothing quite so beautiful as a whole apple, golden and crispy, emerging from your oven. Stuffed with delicious pork mince, currants, and spices, these are very easy to prepare, and never fail to impress. If you don't have an apple corer, a small sharp knife will do the trick.

Preheat the oven to 350°F.

Place the split peas in a small saucepan of water and cook for about 20 minutes or until tender.

Using a small knife, slice a lid from the top of each apple. Set these lids aside (you'll need them later). Using an apple corer or small knife, dig a large cavity in each apple, leaving between ¼ and ⅖ inch around the sides and on the bottom.

Over a low heat, fry the onions with a little fat for 5 to 10 minutes or until they caramelize.

In a mixing bowl, combine the onions, pork mince, spices, and currants. Drain the split peas and add them to the bowl. Season to taste.

Fill each apple with this mixture, then put the lids back on.

Preparation time: 15 minutes
Cooking time: 25 to 30 minutes
Makes: 12 stuffed apples
   (serves 4 to 6)

Ingredients:
⅓ cup split peas
12 medium cooking apples
   (heritage varieties work best)
1 large onion, finely chopped
fat for frying
1 pound pork mince
1 teaspoon dried cinnamon
½ teaspoon dried nutmeg
¼ cup currants
⅓ cup wine vinegar
2¼ tablespoons whole cane or
   brown sugar

Arrange the apples in a large baking dish so that they fit snugly and stand upright. Add ½ cup of water to the base of the dish, then bake the apples for 10 minutes.

Meanwhile, in a small saucepan, combine the butter, vinegar, sugar, and ½ cup of water. Bring to a gentle simmer and stir well.

Remove the apples from the oven. Lift each apple's lid, pour in a little sauce, then replace the lid. Return the tray to the oven and cook for a further 15 minutes or until the apples are soft and lightly brown, but not split or cracked. Serve immediately.

marrow on toast

Bones are filled with a delicious and nourishing fat called "marrow," which occupies the cavities of larger, longer bones in livestock. This was highly prized in hunter-gatherer communities and considered just as valuable as offal and brains (other favorites of our ancestors). Fortunately, you don't need a spear and club to enjoy this dish! The bones are baked and you can very elegantly scoop out the fat and spread it on toast with some fresh parsley and garlic.

Prepare the bones by soaking them for 24 to 48 hours with 1½ tablespoons of sea salt in a bowl of water in the fridge. Change the water once or twice if you can.

When you're ready to cook the bones, preheat the oven to 350°F.

Grease a baking tray with a little cooking fat. Drain the bones and arrange them on the tray. Cook them in the oven for 10 to 15 minutes, or until the marrow is soft and beginning to rise. It should bubble and pop and become a light golden color.

Soaking time: 24 hours
Preparation time: 5 minutes
Cooking time: 15 minutes
Serves: 6 (as an entrée)

Ingredients:
6 beef marrow bones, 1 or 2 inches long
sea salt
1 generous bunch of flat-leaf parsley
2 lemons
6 slices of rye sourdough or other wholegrain bread
6 (or more) large garlic cloves

Meanwhile, finely chop the flat-leaf parsley. Juice both lemons and grate the rind of one of them. Combine the parsley, lemon juice, and lemon rind in a serving bowl and toss well.

Lightly toast the sourdough bread. Serve the marrow on a platter with the parsley salad, toast, and garlic cloves in separate dishes on the side. Each person takes a slice of toast, rubs it with garlic, adds a scoop of marrow fat, then sprinkles some parsley salad on top.

moroccan rabbit hot pot

I envy people who live in the country. I'd love to be able to go out to the back field and catch my own dinner. When an animal is killed this way, you have the assurance that it lived as close to nature as possible, and that its ending was swift. Rabbits are a major threat to our native fauna and flora, so here is a dish that can fill your belly and clear your conscience at the same time. You can often spot young kids selling rabbits for next to nothing on the side of the road in rural areas. Alternatively, you should be able to pick a rabbit up at your local farmers' market. The Moroccan flavors of this dish go well with wet polenta or brown rice and a green salad.

If your butcher hasn't already done it for you, you'll need to cut the rabbit into joints. Cut at each shoulder and hip joint, then slice down the middle of each rabbit's back. Break the resulting bits into casserole-sized pieces that will fit easily into your cooking pot.

Combine the peppercorns, garlic, salt, cinnamon, ginger, and olive oil. Rub the mixture over the meat, then cover and refrigerate for 6 to 24 hours.

Preparation time: 10 minutes
Marinading time: 6 hours
Cooking time: 30 minutes
Serves: 6

Ingredients:
2 fresh rabbits
2 teaspoons black peppercorns
2 cloves garlic, crushed
1 teaspoon sea salt
1½ teaspoons cinnamon
1 teaspoon ground ginger
¾ tablespoon olive oil
8 large, ripe tomatoes, or 2
　cans diced tomatoes
rind of 1 small lemon, cut into
　strips
½ cup sherry, white wine, or
　verjuice
2 cups chicken stock
a few large sprigs of thyme
1 red onion, finely sliced
1 bunch fresh coriander, finely
　chopped

Preheat the oven to 350°F.

In a frying pan over medium heat, fry the meat for a minute or so on each side until it seals.

Put the meat, tomatoes, lemon rind, sherry, stock, thyme, and onion in a heavy-based pot with a tight-fitting lid. Ideally, the ingredients should take up ¾ of the space in the pot. Check that there is enough liquid to just cover the meat, keeping in mind that if you are using fresh tomatoes, they will release a lot of juice.

Put the pot in the oven and cook for 30 minutes, or until the meat is soft. Be sure not to let it overcook, as rabbit meat can become dry and leathery if cooked for too long.

Remove the pot from the oven and season the stew to taste. If you wish to thicken the sauce, remove the meat from the pot and set it aside while you let the sauce simmer, uncovered, for 5 to 10 minutes. Finely chop the coriander or parsley and stir them into the liquid.

Serve each piece of meat with a generous pouring of sauce.

Baked Meatballs with Nutmeg

These meatballs can be whipped up with very little fuss. Serve them as a snack at a party, or combine them with some vegetables for a main meal. They also work well in sandwiches and make an excellent lunchbox snack.

Preheat the oven to 350°F.

In a food processor, combine the meat, onions, nutmeg, egg, breadcrumbs, garlic, parsley, and basil. Pulse the mixture a few times until well combined, then season to taste.

Grease a shallow baking tray with a little olive oil and rub some oil onto your palms to stop them sticking. Roll the mince into golfball-sized balls and arrange them on the tray. Bake for 15 minutes, or until the meatballs are cooked right through and no longer pink.

Preparation time: 15 minutes
Cooking time: 15 minutes
Makes: 24 meatballs

Ingredients:
1 pound minced meat
2 small onions, coarsely chopped
½ clove freshly ground nutmeg or ½ teaspoon dried nutmeg powder
1 egg, lightly whisked
1 ⅔ ounces breadcrumbs (or 1 slice of bread whizzed in the food processor)
2 garlic cloves
½ cup flat-leaf parsley or celery leaves
½ cup basil leaves
olive oil

FRESH FROM THE SEA

"Give me a fish and I will eat for a day.
Teach me to fish and I will eat for a lifetime."

—Chinese proverb

W E ALL KNOW THAT FISH IS GOOD FOR US. Most nutritionists recommend that people eat it several times per week. Seafood is rich in the omega-3 fatty acids DHA and EPA, which are known to benefit heart health, brain development, and metabolic functioning. Seafood is also an excellent source of important minerals, most of which reside in the bones. Smaller varieties of fish such as sardines and mackeral can be eaten whole, while the bones of larger varieties can be added to soups and stews to produce a nourishing and gelatinous broth.

For a long time, however, I had very little knowledge about which fish to choose, or how to cook seafood healthily on a low budget. I purchased the odd salmon steak or whole snapper from my local fishmonger, but felt confused whenever I tried to venture beyond the realm of pan-frying or simple grilling. For a novice, the thought of yourself with a knife in one hand and a squirming little creature in the other can be beyond daunting, and the choices involved in buying fish can be very confusing.

Many people tend to stick to old habits when it comes to cooking fish. It's easy to favor the basic staples such as tuna steak and ignore some of the less popular—but fresher and cheaper—varieties. It never ceases to amuse me that one country's firm favorite may be another country's oddity, and seafood that is overlooked in one part of the world may be highly prized in another. Eels are rarely eaten in the United States but are much loved in Europe. In Australia, carp are considered a pest suitable only for use as a fertilizer, while in some Asian and European countries they are treated as a delicacy. With this in mind, why not try something different the next time you are at the fish market? Find out what is available, fresh, and sustainably caught at your local fish market or fishmonger's. This chapter will cover the basics of choosing fish; cost-effectiveness, freshness, and sustainability. I'm also very much in favor of making the most of every little flipper we purchase—using the bones for stock and the whole fish for soups and stews. Fish were also traditionally pickled, air-dried, or fermented to last through the seasons when it was not available. I have included a few recipes that cater for this (just in case anyone reading this book does not possess a refrigerator).

Lastly, don't be afraid of failure. If it doesn't work, or if your fish goes stale in your fridge before you remember to cook it, simply get out the shovel and bury it in your garden. Rich in iodine and other minerals, a fish buried deep under your vegetable patch will do wonders for whatever you are growing.

BUYING FISH

Price: Since most seafood is still caught in its natural environment, its supply, and consequently its price, is more erratic than that of meat and vegetables. As a general rule, seafood that is in season will be less expensive—which, luckily for frugavores, means that the cheapest seafood is often also the freshest. Wherever possible, choose fish that is in season (your fishmonger will be able to tell you which these are).

Even so, fresh fish can be expensive. But with some traditional peasant know-how, you can obtain good, nutritious fish at a fraction of the cost. If you are going to spend money on seafood, buy the best quality you can afford and use every little morsel. Here are my frugavore tips for your next trip to the fish market:

- Buy the whole fish wherever possible. As well as getting a better price, you'll get the bones as well as the flesh, and the bones can be made into delicious and nourishing soups and stocks.
- Try smaller varieties such as sardines and herring. These are rich in nutrients, but are often a fraction of the cost of larger fish. If you find them at a good price, buy them in bulk and pickle or preserve them.

- Make fish stock. Fish bones, heads, and carcasses are often given away for free (particularly at markets with a high turnover). Stock made from them is rich in vital nutrients, minerals, and iodine, and is extremely cheap and easy to make.

Freshness: Whenever possible, try to buy your fish on the day you want to use it. If you can, shop at a market close to the water's edge; this can be a good assurance that the product is fresh and locally caught.

How can you tell that a fish is fresh? For starters, the eyes should be clear and protruding, with black pupils and transparent corneas. The gills should be a pinky-red color, not brown. Your fishmonger may let you touch the fish: it should be soft and springy, like the rubber on a trampoline. As for the smell, when fish is truly fresh, it should smell only of the ocean, a lovely clean and salty fragrance. If it smells "fishy," don't buy it. The only exceptions to this rule are sharks, skates, and rays, which are best cooked and eaten a few days after they were caught. They contain a chemical called urea, which breaks down after they die, producing ammonia. A fishy smell from them is good—it indicates that the urea is disappearing.

It is always preferable to choose a whole fish and have it filleted than to buy pre-cut fillets, which will lack flavor and probably not be as fresh. If you do buy fillets, look for translucent rather than milky flesh. Fillets that are dry around the edges or show signs of discoloration will most likely be stale.

Sustainability: Traditionally, fish was a staple food for anyone who lived near the sea or had access to a friendly fishmonger. The day's catch would be displayed for locals to pick and choose from, either at the water's edge or at a local market. Choices were fairly easy to make, as people had a clear understanding of what was locally available.

Nowadays, when I visit the supermarket and see a label that says "salmon," "tuna," or "monkfish," my mind draws a blank. Where has this fish come from? Is it healthy? Contaminated with mercury? Sustainably or unsustainably farmed? These questions are reasonable. Buying fish is an ethical issue. Overfishing is now widely acknowledged as the greatest single threat to marine habitats. Over 70 percent of the world's fish stocks are now fully or over-exploited. Further damage is caused by unsustainable modern fish farms and "by-catch" such as turtles and dolphins being killed by modern trawlers.

So what should the frugavore cook be looking for? This is where a good rapport with your local fishmonger comes in handy: frozen fillets from the supermarket will not usually come with any information about where the fish came from, but your fishmonger will be able to tell you which fish are local, fresh, and sustainably fished or farmed.

Sustainable seafood can be sourced either from the wild or from fish farms. If you are buying wild fish, choose fast-growing, highly productive species and avoid fish caught by methods (such as trawling) that damage ocean habitats. If you can't get this information from your fishmonger, try exploring the online resources at the back of this book.

If buying farmed fish, you will want to be sure that they have been farmed using sustainable practices. As a solution to overfishing, farming is not all it's cracked up to be. Just like farms on land, modern aquaculture can produce chemical runoff from antibiotics, pesticides, and detergents. Not surprisingly, fish grown on farms are often not as healthy as wild fish; they have traces of chemicals and antibiotics in their flesh, and contain less muscle tissue and fewer omega-3 fatty acids than their wild counterparts. Carnivorous species, including salmon, tuna, cod, and shrimp, require more food to grow than they produce: it takes roughly three pounds of wild fish to produce one pound of farmed salmon or shrimp.

Truly sustainable farming uses organic feeding practices to farm a wide variety of fish. It involves small, closed aquaculture systems that do not destroy coastal habitats or depend on wild fish as feed. Smaller varieties of fish such as sardines, mackerel, and trout, as well as molluscs such as oysters, mussels, and clams, are best suited to farming, as they do not need wild fish as food.

To find out more about the ecological impact of fishing methods used in your area, get in touch with your local marine conservation group. In the United States, The Natural Resources Defense Council offers information about choosing sustainable seafood online at www.nrdc.org.

Mercury toxicity: There has been concern during recent years about rising levels of mercury in fresh seafood. Mercury and other toxic metals accumulate in larger varieties of fish such as shark, tuna, marlin, and Spanish mackeral. To avoid ingesting mercury, try to purchase the smaller varieties of oilier fish such as sardines, mackerel, and shellfish, which will have accumulated only minimal amounts of mercury.

Cooking fish

It's a real shame that so many of us are unfamiliar with what to do with a fresh fish straight off the boat. We have become so accustomed to buying single fillets or even fish fingers, crumbed, frozen, and boxed. But nothing quite beats the beauty of a whole fish fresh from the hold of a fishing boat. Taking it home with you to clean and fillet yourself is a special experience; you get to see the whole process from start to finish. It is also much more economical, the flavor is supreme, and you have access to all the nutrients contained in the bones, whether you eat them along with the fish (in the case of smaller varieties), or cook them up into a broth.

Some of the recipes in this chapter call for fillets of fish rather than the whole animal. It's still a good idea to buy whole fish for these dishes, as the bones and head are full of valuable minerals and can be used to make stock (the heads are particularly gelatinous). Ask your fishmonger to fillet the fish for you, and take the bones and the heads home in a separate bag.

Cleaning your fish: If you are lucky enough to catch your own fish, you'll need to clean and fillet it as soon as possible after catching it. Use the back of a knife to scrape off the large scales, working from the tail to the head. Then slit the fish open along the length of the belly up to the gills. Remove the gills, empty the cavity of all the guts and scrape away any dark blood. Rinse the fish well before placing it in the fridge or freezer.

Filleting your fish: If your fishmonger won't fillet the fish for you, take the whole fish home and do it yourself. Don't be daunted by this; simply make a cut in the flesh as close to the tail as possible. Hold the knife parallel to the backbone and, pressing firmly, slice along the flesh alongside the backbone until just before the gills. Remove the ribcage and the small line of bones that runs about 1¼ inches in from the thick end of each fillet, then rinse the fish clean of any excess blood or guts.

Steaming: A vegetable or pasta steamer can be used to steam fish, but the best implement is a Japanese bamboo steamer, which you can usually find for a couple of dollars at an Asian grocer or market. Make sure you take your saucepan in with you, or at least its measurements, so that you get the right size.

With the right equipment, steaming is quick and easy. A small fillet (¼ or ¾ inches thick) should steam in less than 5 minutes, and a larger whole fish (2 or 2⅓ inches 9 thick) should take 10 to 12 minutes. Firm-fleshed fish such as salmon, trout, or flounder are the best candidates for steaming.

Baking: This method works best for whole fish. Preheat your oven to 350°F. On a large baking tray, place your scaled and gutted fish and add any herbs or flavorings you fancy. Bake for 20 minutes per 2 pounds.

Pan-frying: With some butter and fresh herbs, this should please even the fussiest of fish-eaters. In Australia we often use fillets of King George whiting, fried with a squeeze of lemon and some freshly chopped parsley, chives, or tarragon.

Poaching: This method requires a little more time, but is well worth the effort. In restaurants they use professional fish steamers (rectangular steel pots where the fish lies on a perforated platform) but I have achieved very good results using just a deep frying pan and poaching the thick fish fillets with stock, a little wine, vinegar, and lemon juice.

Preserves & pickles

Traditionally, anyone living close to the sea would have had all sorts of preserved seafood in their pantry. Traditional methods of air-drying, salting, and fermenting filled the pantry for times when fresh fish was not available. These traditional methods might not be as essential today, when we are not faced with the same periods of scarcity. But these dishes are fun to make and can be very useful to have on hand for weekday sandwiches, hors d'oeuvre, or any of those can't-think-of-what-to-eat situations.

Seaweed

Seaweed is rich in important minerals, particularly iodine, which is a common nutritional deficiency. It also contains a form of gelatin, which has a soothing effect on the intestinal system. All sea vegetables can be added to stocks or used to flavor fish or vegetable dishes. The price of seaweed varies depending on where you shop. Asian supermarkets and grocery stores stock them fairly cheaply, whereas fancy organic stores sell them for much higher prices.

The best varieties to use for home-cooking include kombu, wakame, arame, and dulse flakes. Kombu, arame, and wakame can be added to soups and stews for extra flavor,

minerals, and gelatin. Dulse flakes can be used in place of sea salt to flavor dishes.

oven-baked sardines with oregano

Sardines are one of the most underrated fish. They are always cheap and are surprisingly delicious. Being a small and oily variety of fish, sardines are also full of healthy fats, and their bones (which are edible) are rich in minerals. The first time I made this dish I got goosebumps, they were so good.

Preheat the oven to 350°F.

If you bought the sardines as whole fish (rather than fillets), you will need to fillet them. Slice each sardine along its belly, then butterfly it out (spreading each of the side fillets out) and remove the guts. Chop off the heads, then rinse off any excess blood and guts under cold running water. Discard the head and innards (I usually bury them in the garden; they are excellent fertilizers and will give your herbs extra bite. They do especially well buried under parsley or chives).

Place the fillets, innards-side up, on a baking dish. Drizzle them with lemon juice and sprinkle with sea salt, oregano, garlic, and onion. Add a good dash of olive oil. Put the tray in the oven and cook for 10 minutes, or until the fish are lightly browned but still juicy and soft. Serve with a green salad or on some fresh crusty bread.

Preparation time: 5 minutes
Cooking time: 10 minutes
Serves: 1

Ingredients:
7 small or medium sardines
juice of half a lemon
sea salt
1 small handful fresh oregano,
    finely chopped
1 large clove garlic, crushed
    and finely chopped
1 medium red onion, thinly
    sliced
olive oil

Baked Whole Fish with Tomatoes, Herbs, & Fennel

This is an Australian classic, usually done with one large fresh snapper, although you can also use a large bream or a reef fish such as red emperor. You can chop and change the seasonings depending on what you have handy.

Preheat the oven to 350°F.

Place the whole fish in a large ceramic or stainless steel tray. Finely slice the onion and fennel. Place half of the onion inside the cavity of the fish with some of the fresh thyme and 1 of the bay leaves. Arrange the rest of the onion and all of the fennel around the fish, along with the remaining thyme and the second bay leaf.

Finely slice the tomato and layer it over the onion and fennel, then pour over the white wine.

Rip the bread into tiny pieces or pulse it in a food processor so that it forms neat breadcrumbs. Toss these in a small bowl with the flat-leaf parsley and olive oil and season to taste. Spread the breadcrumb mixture over the top of the fish.

Preparation time: 5 minutes
Cooking time: 30 minutes
Serves: 4

Ingredients:
1 large whole snapper, scaled and gutted
1 large red onion
1 large fennel
¾ tablespoon fresh thyme, finely chopped
4 or 5 large tomatoes
2 bay leaves
¾ cup dry white wine
2 slices crusty sourdough bread
½ cup finely chopped flat-leaf parsley
¾ tablespoon olive oil

Place the dish in the oven and bake for 30 minutes at 350°F. To check whether the fish is ready, pull a portion of the flesh away from the bone. It should be white-colored and separate easily from the bone. Serve immediately with a green leafy salad and some baked potatoes.

FISH PIE

Fish pie has long been a staple for people living near the water's edge. Any sort of fish will do, but I recommend a white-fleshed fish. Before you start, make sure the fillets are free of bones.

Preheat the oven to 350°F.

Finely chop the onion and fennel and fry them over a gentle heat for 15 minutes or until they are soft and lightly browned. Add the mushrooms and cook for a further 10 minutes or until the mushrooms have cooked through and reduced in size.

Scrub the potatoes and put them into a cooking pot. Cover them with water, bring to a boil, and cook for 10 minutes or until soft. Drain them, then either purée them in a food processor or mash them well with a hand masher. Season with salt, pepper, and butter to taste. Add the eggs one at a time, stirring to form a smooth paste.

Pan-fry the fish with a little cooking fat so that it is evenly cooked on both sides. Remove it from the pan and cut it into bite-sized pieces. Stir this through the mushrooms and fennel.

To make the white sauce, place the butter and arrowroot powder in a small saucepan. Stir over high heat until the butter melts and a smooth paste forms. Gradually add 1 cup of

Preparation time: 20 minutes
Cooking time: 25 minutes
Serves: 6

Ingredients:
1 large onion
1 medium fennel
7 ounces mushrooms
⁴/₅ pound potatoes
3 eggs
1 pound white-fleshed fish fillets
2 tablespoons butter
¼ cup arrowroot powder
1 cup flat-leaf parsley, finely chopped
1 handful fresh dill, finely chopped (optional)

hot water in a steady, even stream, stirring continuously. Allow the sauce to simmer and thicken for a further minute.

Stir the sauce through the mushrooms and fish, then add the fresh herbs.

Pour this mixture into an ovenproof pie dish. Spread the mashed potato over the top to form a lid and add a few extra dollops of butter to the top.

Cook in the preheated oven for 25 minutes or until the pie top is golden.

POACHED FISH WITH SABAYON SAUCE

Poaching fish is a wonderful way to keep it tender and moist. You also retain all the important minerals and the herbs from the poaching liquid. Poaching produces delicious results and works well with flathead or any other firm, white-fleshed fish. Sabayon is a traditional French sauce that goes well with any type of seafood.

In a deep frying pan, bring the stock, onion, lemon juice, garlic, and wine to a boil. Reduce the heat and simmer for 3 to 5 minutes so that the liquid reduces a little. Make sure you have enough liquid left to fully cover the fish as it cooks. If you don't, you may need to add some extra.

Add the fish to the pan and cook for 3 minutes on each side, or until the centers of the fillets are no longer pink. When the fish is cooked, remove it from the pan with the sliced onions and place it on a serving dish. Be sure to retain the cooking liquid in the pan.

To make the sabayon sauce, pour the cooking liquid through a sieve.

Put a small or medium bowl inside a saucepan and add enough water to the pot so that the water comes halfway up

Preparation time: 5 minutes
Cooking time: 15 minutes
Serves: 4

Ingredients:
2 or 3 cups fish stock
1 red onion, finely sliced
juice of 1 lemon
2 or 3 cloves garlic, crushed
½ cup dry white wine
1 pound filleted fresh white fish

FOR THE SAUCE:
2 egg yolks
½ cup leftover poaching liquid
salt and pepper
⅔ cup butter, cut into small
   cubes
¾ tablespoon finely chopped
   dill

the sides of the bowl. Turn the heat to its lowest temperature and allow the water to get hot but not boil.

Place the egg yolks in the bowl with half a cup of the strained poaching liquid. Whisk them together well, then add salt and pepper to taste. Continue to whisk, adding the butter as you go, one cube at a time. Keep whisking and adding butter until the sauce thickens and coats the back of a wooden spoon.

When the sauce has reached the desired consistency, take the pan off the heat and continue whisking for about 30 seconds as it cools. Pour the sauce into a warmed jug and stir through the finely chopped dill before serving.

minced fish cakes

These little fish balls are delicious with sour cream and dill, or sauerkraut and a baked potato. If you make them small enough, they are also a wonderful hors d'oeuvre, or packed in a lunchbox and taken to school or work.

I'm not saying you should deliberately buy fish that isn't fresh—but keep in mind that you don't have to buy the most expensive fish for these cakes. Any white-fleshed fish that is free of bones will do.

Preheat the oven to 350°F. Generously grease a large steel or glass baking tray with butter.

Boil the potato until it's soft, then mash it.

Combine the mashed potato with all the other ingredients in a food processor or by pounding them together in a mixing bowl with a potato masher.

Roll the mixture into 1-inch round balls with your hands. Place each ball on the baking tray and bake for 15 minutes in the preheated oven. Serve hot or cold.

Preparation time: 5 minutes
Cooking time: 15 minutes
Serves: 4

Ingredients:
1 small potato
1 pound filleted fresh fish
⅓ cup plain full-cream yogurt
½ teaspoon freshly ground nutmeg
¼ cup finely chopped parsley
1 large clove garlic, crushed and finely sliced
½ teaspoon finely chopped chili flakes (optional)
1 teaspoon butter
salt and pepper

FISH SOUP

Fish soup depends on a good hearty fish stock, made from bones or offcuts and some fresh vegetables, herbs, and seasonings. Fish stock is highly nutritious—it is rich in iodine, minerals, and gelatin—and is also very cost-effective to make at home. The heads, tails, and bones of larger fish, sometimes referred to as "scraps" by fishermen, often sell for next to nothing at fish markets.

Traditionally, the contents of fish soup would vary depending on the day's catch and the local region's preferred flavors. Bouillabaisse is a superb French Mediterranean dish made with tomatoes and plenty of saffron. In Greece, a variation involves rice, eggs, and lemon. In Portugal and Spain, spicy flavors such as red peppers, paprika, and chili dominate. The best soups are made with what is available locally, with the seasonings and spices that you enjoy the most.

FISH STOCK

Many fishmongers give away carcasses and fish heads for next to nothing, so fish stock can be ridiculously cost-effective. Fish heads are particularly nutritious, as they are rich in gelatin and fatty tissue and produce a deliciously rich and healthy stock. Avoid using oily fish such as salmon or mackerel when making stock. Their fragile oils will oxidize and create a very smelly stock.

Place all the ingredients in a large stockpot and fill it with cold water. Slowly bring it to a boil, then reduce the heat and gently simmer, partially covered, for 30 to 40 minutes, or until the liquid is thick and rich. Check the stock occasionally and crush the bones as they cook so that they break and drop to the bottom of the pot.

When the stock is nice and thick, drain the liquid from the solids and store it in the fridge.

After stock-making, the bones and vegetables can be buried in the garden. Fish bones are an excellent source of iodine and other minerals. If you are worried about dogs or birds digging them up before they have a chance to break down, just place a few rocks over the spot where they are buried.

Preparation time: 5 minutes
Cooking time: 30 to 40 minutes
Makes: 4 to 5⅓ quarts

Ingredients:
2¼ pounds fish carcasses, heads, tails, bones, and other offcuts
¼ cup apple-cider or wine vinegar
2 bay leaves
1 handful fresh or dried herbs, such as thyme or rosemary
1 large onion
2 or 3 sticks celery, including the leaves
2 large carrots, coarsely chopped

Optional extras:
a few strips seaweed, such as arame or dulse
1 or 2 leeks, coarsely chopped
1 fresh knob of ginger
A dash of white wine

FISH BROTH WITH Lemon & RICE

This recipe was born when there seemed to be *literally* nothing to eat in my kitchen. With only some frozen fish stock, a packet of rice, and some eggs from the henhouse, I whipped this up for a late breakfast on a Saturday morning. I was surprised how very good it was, and it's now a regular part of my repertoire, with the addition of some lemon juice and parsley. I cook it whenever I want something simple, quick, light, and nourishing.

Bring the fish stock to a boil in a large pot, then reduce the heat and let it simmer gently. Add the rice and garlic and cook for 30 minutes, or until the rice is soft.

In a large bowl, whisk together the lemon juice and eggs until they are light and fluffy (an electric beater makes this easier, but a hand whisk does the job too).

Allow the fish stock to cool until it is still hot but no longer boiling. Pour it through a sieve into the bowl with the egg mixture, leaving the rice in the pot.

Whisk together the eggs and stock for 30 seconds, and then return them to the pot with the rice. Season with salt and pepper; the soup is now ready to serve. Ladle it into individual bowls and garnish with the parsley and olive oil.

Preparation time: 5 minutes
Cooking time: 30 minutes
Serves: 3

Ingredients:
1 quart fish stock
¾ cup rice
2 cloves garlic (optional)
juice of 2 lemons
3 eggs
salt and pepper
1 generous handful flat-leaf
   parsley, finely chopped
olive oil

Beautiful Bouillabaisse

This is such an easy and delicious dish to put together. If you buy whole fish, you can fillet them yourself and use the bones and heads for stock and the fillets for the bouillabaisse. Alternatively, buy them filleted and make your stock in advance, and you'll be able to whip up this beautiful dish in less than an hour.

Finely chop the onions. Crush and finely chop the garlic. Cut the potatoes into cubes or quarters, and slice the fennel into thin slivers.

In a large pot, combine the onions, garlic, fish stock, canned tomatoes, wine, and orange peel. Bring to a boil, then reduce to a gentle simmer. Add the potatoes, fennel, and saffron and cook for a further 10 minutes, or until the potatoes are soft.

While the stock simmers, prepare your fish. Cut the fillets into bite-sized pieces. Clean the mussels and have them ready in a bowl next to the stove.

Preparation time: 10 minutes
Cooking time: 30 minutes
Serves: 6 (generously)

Ingredients:
1 medium onion
2 or 3 garlic cloves
3 or 4 medium potatoes
2 medium fennel
3¾ quarts fish stock
3 cans (16 ounces each) diced tomatoes
½ cup dry white wine
1 long curl orange peel
½ teaspoon saffron threads
1¾ pounds white-fleshed fish such as blue grenadier, rockling, or cod
2¼ pounds mussels or other crustaceans
2 teaspoons finely chopped fresh thyme
1 large handful fresh parsley, finely chopped

Increase the heat under the pot and bring the stock to a strong simmer. Check the liquid levels; the broth should be thick and gelatinous and not too watery before you add the fish (simmer it a little longer if you need to). Add the fish to the pot and let it cook. Stir well and season to taste.

Add the mussels 5 minutes before serving. Cover the pot and increase the heat to a steady boil for 3 minutes. Serve in individual bowls with generous scatterings of parsley and a dash of olive oil.

bermuda fish chowder

Bermudians love rum. They also love pepper and cloves, and this chowder combines these flavors sublimely. This dish is rich and gelatinous and requires long, slow cooking. Enjoy it with some crusty bread—and an additional dash of rum or pepper, if you want an extra kick.

Finely chop the onion and crush and finely chop the garlic. Peel and dice the potatoes and carrots and thinly slice the celery. Grind the peppercorns and cloves with a mortar and pestle.

In a large pot, combine the stock, bay leaves, onions, garlic, celery, carrots, potatoes, thyme, tomatoes, tomato paste, and Worcestershire sauce. Bring the liquid to a boil. Add the ground peppercorns and cloves, then reduce the heat so that the stock simmers and gurgles. Let it simmer, partially covered, for 45 to 50 minutes.

At the end of this time, cut the fish fillets into bite-sized pieces and add them to the pot along with the rum. Simmer, partially covered, for a further hour, or until the stock is thick and gelatinous and the vegetables are tender and soft.

Preparation time: 20 minutes
Cooking time: 2 hours
Serves: 8 to 10

Ingredients:
2 onions
3 garlic cloves
8 small to medium potatoes
(about 1¾ pounds)
6 medium carrots
1 stalk celery
2 teaspoons whole
peppercorns
1 heaped teaspoon whole
cloves
4 quarts fish stock
6 bay leaves
¾ teaspoon fresh thyme
4 cans (16 ounces each)
chopped tomatoes
1½ tablespoons tomato paste
1½ tablespoons Worcestershire
sauce
¼ cup rum

Ingredients continued ...

1½ pounds fish fillets
salt and pepper
1 generous bunch flat-leaf
   parsley, finely chopped
olive oil

Season with salt and pepper to taste, then serve with a generous sprinkling of parsely and a good dash of olive oil for each bowl.

salmon gravlax (cured salmon)

Gravlax was traditionally made by Scandinavian fishing communities using a combination of salt and herbs to preserve the fish for long periods. The fish was packed with salt and buried in the ground so that it lightly fermented. The word "gravlax" comes from a Scandinavian word for "grave," while "laks" means "salmon." Now, there is no need for a garden burial; you can easily prepare this dish in your kitchen. Gravlax goes beautifully with sour cream, dill, pickles, and a dark rye bread.

Have ready a large ceramic or glass dish, big enough for the fish to sit comfortably without spilling over the sides.

Remove the small bones from the thickest part of the fillet using tweezers. Combine the sugar, salt, dill, and vodka in a small bowl and rub the mixture over the sides of the fish. Sprinkle some of the mixture into the bottom of a large dish and place the fish on top. Pack the rest of the mixture around the sides of the fish.

Cover the fish with a sheet of greaseproof paper and weigh the paper down using a bag of sugar or dried beans or something similar. This will ensure the fish is packed in tightly and that water is able to seep out as it is drawn from the fish.

Preparation time: 10 minutes
Refrigeration time: 1 or 2 days

Ingredients:
1 large salmon (at least 2¼ pounds)
2 cups brown sugar
2½ cups rock salt
¾ cup freshly chopped dill
¼ cup vodka

Place the dish in the fridge for at least 12 hours. I usually refrigerate mine for about 24 hours. The longer it's in there, the stronger the flavor will be. Every 12 hours, turn the fillet over and drain out any excess water.

When it's ready, thinly slice the fillet and serve with bread, sour cream, and fresh dill. It will last for a good couple of days in the fridge.

PICKLED FISH

At our local organic health-food store, they sell neatly pack-aged sardines in olive oil for seven dollars per can. Each can contains three little sardines and a bit of olive oil. When I told my fishmonger I'd been buying these, he had a good laugh. For the price of one can you could buy a whole 2 pounds of fresh sardines. He thought it was an absolute joke that some-one would pay so much for what she could easily make herself at home.

You can use any small, oily fish for this recipe, including pil-chards, mackerel, tommy rough, or herring. If you keep them in the fridge, they will last for a period of months. Pickled fish can be thinly sliced and added to sandwiches or salads. I like to make a dip with finely chopped pickled fish, sour cream, dill, and finely chopped cucumber. It's delicious with crackers and a glass of wine.

If you bought the sardines whole, you will need to fillet them. Slice each sardine along its belly and remove the guts. Remove the spine from the center of each fish. Chop off the heads and fold open the flesh. Rinse off any excess blood or guts.

Preparation time: 20 minutes
Refrigeration time: at least
    24 hours
Makes: 2 jars

Ingredients:
1 pound fresh sardines, whole
    or filleted
sea salt
1¼ cups wine vinegar
olive oil
1 small red onion, thinly sliced
1 generous bunch fresh herbs
    (thyme and oregano are my
    favorites)
A few slices fresh chili
    (optional)

Note:
See the "Preserves" chapter
for instructions for sterilizing
glass jars.

Sprinkle salt over the bottom of a ceramic dish. Layer the sardines in the dish, sprinkling salt over each sardine as you go. Ideally there should be 2 or 3 layers of sardines in the dish, packed in tightly. On the top layer, add an extra sprinkling of salt to reduce the fish's exposure to the air. Place the dish in the refrigerator and leave it for at least 8 to 12 hours if the fish are small; if you are using larger fish (such as tommy rough or herring), I recommend leaving them for closer to 24 hours. If you leave them for this long, check on them once or twice, draining off any excess fluid and adding a little extra salt.

When you've left them for the desired amount of time, transfer the fish to a plate and dust off any excess salt. Rinse out the dish and wipe it clean.

Return the sardines to the dish and cover each layer with vinegar. Try to ensure that the fish are packed in tightly and fully immersed in the vinegar with minimal exposure to the air. Place them in the refrigerator for a further 3 to 6 hours. Again, if you are using a larger species of fish, leave them for a little longer.

Remove the dish from the refrigerator and drain off any excess fluid.

Finely chop the onion and herbs. Place the sardines one by one into the sterilized glass jars. Sprinkle them with herbs and onion and cover them with olive oil. It's very important to make sure that there are no air bubbles. I find it easiest to push the sardines up against the vertical walls of the jar, rather than flat on the bottom. The fish should be completely immersed in the oil, so that they have no contact with the outside air. Screw the lid on tightly and keep them in a cool cellar or the refrigerator. They will be ready to eat in 2 days.

GOOD GRAINS

"How can a nation be great if its bread tastes like Kleenex?"

—Julia Child

I LOVE COOKING WITH GRAINS, BE IT AN oatmeal slice filled with fruit or a freshly baked loaf of sourdough bread piping hot out of the oven. There are very few weekends when I don't have something rising or fermenting in my kitchen. Grain-based foods, when sourced and prepared correctly, can be highly nutritious and seriously good to eat.

Without the right preparation, however, grains lose most of their nutrients and fiber and can be difficult to digest. So if you are one of the many people who don't really like bread, porridge, or oatmeal, or react badly to the soft and fluffy loaves from the supermarket, keep reading. This section covers some time-honored staples—oats, polenta, rice, and bread. The recipes all employ traditional cooking techniques to deliver a tasty and highly nutritious product with minimal fuss. I hope they will make you fall in love with traditional bread-making and old-fashioned oatmeal slice, just like I did.

WHOLEGRAINS

A wholegrain consists of a husk, germ, and endosperm. It is a highly nutritious ingredient, full of fiber and many essential nutrients. But when a wholegrain is processed to make an industrially produced bread or cereal, the germ and endosperm are removed—along with most of the fiber and protein and a significant proportion of the vitamins. The result is very different from traditional peasant-style breads and gruels.

Wholegrains are quite volatile and very sensitive to heat and light. When you are preparing them at home, they need to be stored in a fridge or freezer, or used very quickly to prevent them from rotting. In comparison, refined grains, which are stripped of many of their nutrients, can last for months on the supermarket shelf. You might see a "wholegrain" label on your packet of cereal or loaf of bread. Don't be fooled. These have usually been treated with a range of fungal inhibitors, additives, and preservatives to prevent them from rotting, so they don't have the natural goodness of a traditional wholegrain product.

Food companies are well aware of the many nutrients that are lost during grain refinement. To compensate, breads and cereals are often fortified with iron, B vitamins, and fiber. The problem with fortification is that many of these nutrients can only be assimilated with the help of naturally present enzymes and co-enzymes that are not present when foods are fortified.

Peasant-style cooking methods, by contrast, not only used the whole grain; they also employed preparation methods such as soaking, sprouting, and fermenting that were designed to enhance the available nutrients and break down any indigestible components, such as phytates, that are naturally present in grains. To my mind, everyone should be preparing their own grains at home, or looking for a good baker to do the same job.

BREAD

Do you love your bread? Do you relish every nuance of taste and texture? Does it digest well? And does it fill you up? A good loaf, rich in wholegrains and leavened for a period of days, should be filling and satisfying, giving you plenty of energy and vigor and keeping you going between meals.

Real bread, like our great-grandparents ate, was made from little other than flour and water. Commercial yeast was only developed during the late nineteenth century. Before this, bread-makers had to rely on the naturally occurring yeasts and lactobacilli found in flour to get their loaves to rise.

Peasant-style bread involves a traditional leavening process. A "starter" is made using flour and water, which turn into a bubbling concoction of live bacteria and yeast. You use this to ferment your loaf and to make your dough rise. Sourdough bread has its characteristic sour flavor as a result. During the leavening process, the healthy lactobacilli in the starter create an acidic environment, which ferments or pre-digests the proteins in the flour. Phytic acid, an "antinutrient" that inhibits the absorption of calcium, iron, magnesium, and zinc, is broken down during this process, as are protease inhibitors, which interfere with the digestion of proteins such as gluten. Anecdotally, many people who cannot tolerate commercially made bread find they have no trouble digesting traditionally made loaves.

Sourdough bread also has a lower glycaemic index than conventionally made bread. It is usually thicker and heavier, which means you feel full after only a few slices. If you buy ready-made bread, keep in mind that sourdough is much better value for money; in nutritional terms, one loaf of sourdough is equivalent to two loaves of conventional bread—it will fill you up more quickly and contains a greater range of important nutrients.

Bread-making is not hard. Once you get into a rhythm, it takes just ten or fifteen minutes in the evening. Then you leave the dough to rest overnight, switch the oven on in the morning, and allow your darling loaves to rise and bake. Easy. If you don't fancy doing this often, you can bake vigorously for a period of a week, then freeze your loaves in bulk. They do not dry out in the freezer, and in fact taste beautifully fresh if you defrost them slowly in a warm oven.

If you are really averse to bread-making and are thinking of skipping this section altogether, that's fine. Just make sure you find a good baker—one who uses organic flour with no odd-sounding additives to make their own sourdough loaves. That is the next best thing to baking your own. But I might add, every time I make bread I feel it should be a cause for celebration. Yeast and bacteria are everywhere—in our hair, on our skin, and in the air we breathe. Sourdough fermentation takes advantage of this, tapping into these naturally present ingredients in your home. Each loaf of bread will be different, reflecting the scents of your kitchen, the air of your neighborhood, and the touch of your fingers.

Choosing a flour

Flour is a volatile food product. It needs to be used quickly, or stored in a fridge or freezer. To avoid it going rancid, many home bakers like to buy their flour as whole wheat (also called "wheat berries") and use a flour mill to grind it themselves. If you are not grinding it yourself, look for stone-ground, organic varieties of wheat. Non-organic flours often do not work in sourdough bread-making because they contain various preservatives and mold inhibitors, which can prevent the fermentation process from taking place.

Commercial "plain flour" is made from wheat that has been specially bred to produce a higher yield and a more glutinous loaf. Many people therefore prefer traditional grains such as rye, spelt, and kamut for both taste and nutritional reasons. If you are a first-time bread-maker, however, I suggest that you use at least 50 percent plain or spelt flour to ensure a good rise in your first couple of loaves. You can experiment with different types of flour once you're a bit more confident. When it comes to your starter, I've found that rye flour works exceptionally well; try to use at least 50 percent rye flour in your starter if you can. Generally speaking, I use a variety of different flours and mix and match as I go. Plain flour is usually a lot cheaper than spelt. If you want to save on costs, you can also buy your flour in bulk and store it in the freezer.

BREAD-MAKERS

I have never used a bread-maker and don't believe they are necessary to make good bread. Some of my sourdough-minded friends do have bread-makers and have adapted these recipes for their machines with great success.

PREPARE YOUR KITCHEN

Make sure you have a clean kitchen that hasn't been cleaned with chemical cleaning products. A sterilized workbench will kill the healthy micro-organizms in the starter. Wash your hands thoroughly before you start and keep your fingernails short. You can also soak your hands in a 50:50 vinegar and water solution for ten minutes before you start.

MAKING YOUR STARTER

The starter is a beautiful and diverse living product, a combination of wild yeast, fungi, and several strains of lactobacilli. It goes to work on the wheat flour, digesting the peptides and breaking down the gluten and phytic acid that are naturally present in wheat.

Making a starter is easy. Get some fresh organic rye flour and filtered, non-chlorinated tap water and you are good to go. The chlorine in tap water will kill many of the micro-organizms required for fermentation. If you don't have access to filtered water, boil your tap water for at least ten minutes and allow it to cool before using it. When following the steps below, keep in mind that the exact time it takes for your starter to develop will depend on your kitchen environment, temperature, and choice of flour.

Day 1: In a small bowl, combine ¼ cup flour with ¼ cup filtered water. Stir well to form a smooth paste. Cover the bowl with a tea towel and leave it on the kitchen counter. If you have a chance, give the starter an additional stir later in the day.

Day 2: In the morning, add a heaped teaspoon of flour and a similar quantity of water to the bowl and stir well. The idea is to keep the same consistency, and the same proportion of flour to water, as you achieved on the first day. Repeat this process again in the afternoon if you can.

Day 3: Now you can start feeding your starter a bit more. In the morning, add two heaped teaspoons of flour and an equal quantity of water and stir well. Repeat in the afternoon if possible.

Day 4: Repeat as per day three.

Day 5: You should be noticing some good bubbles by now. The next step involves a little guesswork. You want to add a quantity of flour equivalent to half the amount of starter. So if it looks like you have 1 cup of starter in your bowl, add ½ cup flour, plus enough water to maintain the same smooth consistency. In the afternoon, add ¼ cup flour and an equal quantity of water for some extra nourishment. Stir well.

Day 6: Your starter should be ready by now, but you can leave it another day if you like. The bubbles should be big and rich, just like honeycomb. If this is the case, you can start your bread-making! If there are still no bubbles, throw it out (or bury it in your compost) and start again.

Maintaining your starter

For good, flavorsome, well-risen loaves, you need to keep the sourdough culture bubbly and active (by giving it a lot of attention, in other words). In warm climates (between 68 and 86°F) it will grow quickly, bubble a lot, and require a lot of nourishment. In cooler temperatures, it will slow down and hibernate.

If you are going away for a week, or won't be baking for a while, you can keep your friend in an airtight container in the refrigerator and it will lie dormant. You'll just need to build it up again with regular feedings and give it some warm air before your next bread-making session. A layer of light brown liquid (known as "hooch") may form on the top of the starter while it is in the fridge. Simply skim this off and discard it before getting started.

The starter likes to be constantly fed and used, so try to feed it every day. At the minimum, add ¼ cup flour and an equal quantity of water. If you really want to build it up, add half of its bulk in flour and an equal amount of water. Give it a good stir each time you feed it. If you bake every day, you can replenish your starter each time you use it. On days when you are not baking, just add a little flour and water to the starter to feed it and keep it happy. I often go on a baking blitz: I'll bake several loaves in one go and freeze every second loaf I make. You can keep your starter in the fridge from Monday through Wednesday,

then take it out on Thursday and start building it up for a baking session on Saturday. If it has been in the fridge it will take a few days of feeding and warmth to get back to its peak of liveliness.

Basic Sourdough Bread

This recipe is for a basic sourdough—simple, wholesome and delicious. Once you've got this down pat, you can experiment with other flavors.

In a large bowl, combine the starter with a little more than 3 cups of the flour. Add just enough filtered water (usually no more than 1½ cups) to form a smooth paste, similar in consistency to the starter. The water quantity will vary depending on the moisture content of your starter. Stir well, then cover the bowl with a tea towel and leave it to sit at room temperature for 12 hours or overnight. It should be bubbling by the morning.

Now you're ready to make your bread. Before you get your hands dirty, have your equipment ready to go. You'll need a large mixing bowl, wide but not too deep. You'll also need a large cutting board—or clear and clean your kitchen counter so that you have room to knead your loaf directly on it. Have your flour ready and a jug of fresh, filtered water on hand, just in case you need extra. Dust your bread tins with polenta; this will prevent the bread from sticking. Lastly, clear plenty of room: things can get messy! As you work, use the olive oil to grease your hands and prevent stickiness.

Preparation time: 15 minutes
Rising time: 24 hours
Cooking time: 1 hour
Makes: One 9¾-inch loaf

Ingredients:
1 cup starter, alive and bubbling
8 cups flour (spelt or plain)
filtered water
1 handful polenta
olive oil
sea salt

Add the sea salt and 4 cups of the flour to the bubbling bread mixture. Mix well so that breadcrumbs form, gradually adding a little more water if necessary to maintain a smooth consistency. If the dough gets too sticky, add a little more flour, but no more than 1 more cup.

Work the dough into a round ball. Place it on your chopping block or bench and start kneading, using firm but gentle wrist motions. This is the only time you'll need to knead your loaf, so give it a good 10 to 15 minutes of your attention. The dough should feel soft and springy. If you poke it, it should slowly spring back to its original form. If it feels like hard pastry, either you haven't kneaded it enough or your starter wasn't active enough. If it's the latter, your loaf might not work.

Once you're done kneading, place your ball of dough into the baking tin. The loaf will double in size, so make sure there's room in the tin to accommodate this.

Next, you'll need to find a place for your tin to sit while the dough rises, preferably somewhere around 17 or 18°C and protected from drafts. I usually keep mine in the unheated oven, where it's safe from flies and moths. Leave your dough to sit for 3 to 4 hours, or until risen significantly.

The next morning, place your loaf in the oven (if it isn't already there) and *then* turn the oven on to 350°F. As the oven slowly heats up, it will add an extra rise to your dough. Most modern ovens will take about 10 to 15 minutes to get to the desired temperature. Once the oven reaches 350°F, bake the bread for 45 minutes. So if you turn the oven on at 7.30 a.m., you can take your loaf out at 8.30 a.m.

A perfectly baked loaf will be well risen and come away from the tin at the sides. To remove it from the tin you should only need to tip it upside down. To check if your loaf is ready, tap it gently; if you hear a nice hollow sound, the loaf is well cooked.

Once it's out of the tin, let your bread rest on the counter for at least 10 or 15 minutes before you slice and eat it.

sourdough fruit loaf or fruit buns

In a large bowl, combine the starter with 3 cups of the flour. Gradually add water, stirring as you go, until it forms a smooth paste, similar in consistency to the original starter. Cover the bowl with a tea towel and leave it to sit at room temperature for 12 hours or overnight. It should be bubbling by the morning.

When you're ready to start baking, gather your equipment and ingredients and dust your baking tin with polenta.

Add 3 cups of the flour to the bubbling mixture, stirring with a wooden spoon.

Add the cinnamon, ginger, nutmeg, dates, currants, sugar, and vinegar. Juice the orange and add half the juice to the mixture (retain the rest of the juice for other cooking, or drink it later). Finely grate the orange rind and add it, too. Stir the dough well, adding a little water if necessary to maintain the smooth consistency.

Grease your hands with a little olive oil. Gradually add the last cup of flour to the dough, while kneading it in the bowl with your hands. Knead the dough in the bowl until it becomes thick and heavy and can easily be taken out as a single ball, then place it on your countertop and knead hard for 10 to 15 minutes. It should feel springy and light, and not in any way

Preparation time: 15 minutes
Resting time: 12 hours or overnight
Cooking time: 1 hour
Makes: One 9¾-inch loaf, or one tray of 16 buns

Ingredients:
1 cup bubbling starter
7 cups flour (spelt or plain)
filtered water
1 handful polenta
1 teaspoon sea salt
2 teaspoons cinnamon
2 teaspoons ginger
1 teaspoon nutmeg
1 cup chopped dates
1 cup currants
1 cup rapadura sugar
2 teaspoons apple-cider vinegar
1 orange
olive oil

dense or hard like pastry. If you press your fingers into the dough, it should slowly spring back into shape.

When you have finished kneading, place your loaf in the bread tin. Or, if you are making buns, roll the dough into small balls and arrange them on a baking tray. The dough will double in size, so be sure to leave enough room in the tin, or enough space between buns, to accommodate this.

Place the tin or tray in the unheated oven and leave it to rest for 2 to 3 hours. Once the dough has rested, turn the oven on to 350°F. When the oven reaches 350°F, bake the bread for a further 45 minutes, or 30 minutes if you are making buns. If you're baking a loaf, you should only need to tap on it to remove it from the tin. Once the bread is out of the oven, let it rest on the counter for at least 10 or 15 minutes before you slice and eat it.

Porridge

People who claim that they don't like porridge have not tasted the real thing. Soaking the oats before you cook them imparts a delicious, tangy flavor. Porridge can be soaked overnight or even for a period of days. The mixture won't go bad, and the texture will become softer and more delicious. Once you get into the habit, putting the oats on to soak before you go to bed becomes second nature and makes for a delicious and nourishing breakfast. I love my porridge drizzled with butter or smothered in cream. You can also sprinkle some fresh fruit or desiccated coconut over the top. Instead of buying artificially flavored oats, raid your pantry, fridge, or backyard; dried coconut, fresh fruit, and plain yogurt are all excellent additions.

Put the oats, vinegar (or lemon juice), and salt in a saucepan, cover them with water, and leave them to soak overnight.

In the morning, place the saucepan on a low-to-medium heat and cook for 5 minutes or until the porridge is smooth and creamy. Remove from the heat and add any flavorings you like.

Soaking time: overnight
Cooking time: 5 to 10 minutes
Serves: 2

Ingredients:
1 cup rolled oats (you can also try rye, spelt, or barley oats)
½ teaspoon apple-cider vinegar or the juice of half a lemon
½ teaspoon sea salt

Variations:
¾ tablespoon of plain yogurt added to the soaking oats will give your porridge a smoother, creamier texture.

Before serving, try adding:
- ½ cup finely chopped dates or dried currants
- ½ cup desiccated coconut
- 1 apple, pear, or peach, finely chopped
- honey or maple syrup and some fresh cream or yogurt

Note:

Try to buy organic oats wherever possible. They are generally only marginally more expensive than conventional oats and are well worth the extra money. "Quick cooking" oats have been finely chopped; they are still nutritionally good for you, but whole oats are better. Oats are known to be good for the heart, as they are rich in indigestible carbohydrates called beta glucans that have been shown to lower blood cholesterol. They also contain a number of phenolic compounds that have been shown to have antioxidant effects. Neither oats nor barley have any gluten-producing proteins, but some people who suffer from gluten-sensitivity can still have a reaction to oatmeal products.

porridge cake

Ever wonder what to do with the leftover porridge stuck to the bottom of your saucepan? You could always give it to your chickens (chickens love leftover porridge; it's like Christmas dinner for them). But after doing this one too many times, I had an idea: soft-centered, beautifully crisp porridge cake. Excellent with a dollop of cream or yogurt, or a cup of tea in the late afternoon … "high tea" in true frugavore style, turning leftovers from breakfast into an afternoon treat. Easy, frugal—and truly delicious.

Preheat your oven to 350°F and grease two 9¾-inch cake tins. Combine all the ingredients in a food processor or a large bowl and mix well. Place the mixture in the greased tins and bake for 50 to 60 minutes. The filling will remain moist but the top will turn golden and crispy.

Serve with cream and a drizzle of maple syrup.

Preparation time: 15 minutes
Cooking time: 60 minutes
Serves: 6

Ingredients:
2 cups leftover porridge (or a combination of leftover and fresh porridge)
3 eggs
½ cup coconut fat or lard
¾ cup brown sugar or 1 cup whole cane sugar
juice and rind of 1 lemon
¾ cup desiccated coconut
a few drops vanilla essence
1 teaspoon dried cinnamon
½ teaspoon dried ginger

oatmeal slice

If I had to choose one dish to embody my entire university experience, it would be this one. I'd often put this slice together late at night, then spoon it into a pan in the morning and bake it while I had my shower. Piping hot oatmeal slice would accompany me either between my knees as I drove to work, or at the bottom of my backpack as I rode my bike to class. It was the perfect student dish: incredibly delicious, incredibly easy, and incredibly cheap. You can also use the recipe for stewed pears (see page 300) as a topping, instead of fresh fruit.

Using a food processor, combine the oats, ground almonds, vinegar, oil, sugar, treacle, cinnamon, and flour. Gradually add water to the mixture until it forms a smooth paste, then leave it to soak overnight.

In the morning, preheat your oven to 350°F and grease a baking tray (I use a rectangular ceramic or stainless steel tray, about 11 inches by 7 inches).

Preparation time: 10 minutes
Cooking time: 55 minutes
Serves: 8

Ingredients:
2½ cups rolled oats
¼ cup ground almonds
1 teaspoon apple-cider vinegar
½ cup coconut oil or lard
½ cup whole cane or brown sugar
¾ to 1½ tablespoons treacle
1 teaspoon cinnamon
¾ cup flour
¼ teaspoon baking soda
2 eggs
1 cup desiccated coconut
2 pieces of fruit, for instance 1 apple and 1 peach

Add the eggs, baking soda, and coconut to the oats mixture and stir well. Pour the mixture into your baking tray. It should be between ¾ and 1½ centimetres thick. Thinly slice the fresh fruit and arrange it on top of the oatmeal.

Bake the slice for 45 to 50 minutes, or until the top is lightly browned and a skewer comes out clean.

oatmeal pastry

This recipe makes a fluffy and delicious pastry, which you can use for sweet or savory dishes (I use this recipe elsewhere in this book). You can also experiment with different fats to produce different flavors and textures. Duck fat is my favorite, but butter, lard, and chicken fat all work well.

Combine all the ingredients in a food processor. Blend them until they are well combined; the oatmeal should be finely ground and the mixture sticky and gooey. Add a few teaspoons of cold water if necessary. Transfer the mixture to a bowl and leave it to rest overnight at room temperature.

The next day, grease a baking tin and, if you'll be baking the pastry now, preheat your oven to 350°F. If your pastry is too warm and sloppy to work with, place it in the refrigerator and allow it to set for 10 minutes.

Use your hands to fold the pastry out into the baking tin. The pastry should be supple enough for you to shape it by hand; you shouldn't need to add any more flour. If it starts to stick to you, you can grease your hands with a little olive oil and start again.

Preparation time: 15 minutes
Resting time: overnight
Makes: 1 quantity (enough to make a pastry base for a 12-inch pie)

Ingredients:
½ cup yogurt
1¼ cups flour
½ cup rolled oats
⅜ cup butter or cooking fat
¼ teaspoon sea salt
½ tablespoon brown or whole cane sugar (optional, for sweet dishes)
juice and rind of one lemon (optional, for sweet dishes)

If you prefer a smoother, thinner pastry, you can roll out the pastry between two sheets of grease-proof paper and spread it with a rolling pin. Place it on the baking tray so that it sits flat.

To bake blind:

Some pie recipes call for you to bake the pastry "blind." This simply involves baking the pastry by itself for a short time before adding your pie filling. Once you have put the pastry in the tin, cover it with greaseproof paper and fill it with enough rice to weigh the paper down (you can buy some cheap rice for this purpose and use it more than once). Bake it at 350°F for 15 minutes, then remove it from the oven and take out the rice and greaseproof paper. Return the pastry to the oven and bake it for a further 5 minutes. Allow the pastry to cool to room temperature while you prepare the filling.

To make an egg glaze:

You can make a glaze for the pie top by whisking 1 egg in a small bowl with a fork. Pour half of the whisked egg onto the pastry and spread it with a pastry brush or the back of a spoon, then bake as normal.

oatmeal pikelets

These are a wonderful breakfast to wake up to on a Saturday morning. I like to serve them with plain yogurt and a drizzle of maple syrup, but they also go nicely with fresh jam, honey, or marmalade.

Combine the flour, oats, yogurt, and milk and stir to form a smooth paste. Cover the bowl with a tea towel and leave it to sit overnight at room temperature.

The next morning, add the sugar, baking soda, lemon juice, lemon rind, cinnamon, and eggs and mix well.

Heat a frying pan over medium heat and add a little cooking fat. Place a small ladleful of the mixture into the pan—enough to spread to about 2 inches wide. Repeat in the remaining space around the frying pan.

When the pikelets are lightly browned on the underside, flip them over. Cook them for a further minute or two on the second side, then remove them from the pan and cook your next batch.

Serve with plain yogurt and maple syrup (my favorite combination), or whatever condiments you prefer.

Preparation time: 15 minutes
Refrigeration time: overnight
Cooking time: 10 minutes
Makes: 20 pikelets

Ingredients:
1 cup flour
½ cup rolled oats
½ cup yogurt
1⅛ cup milk
¼ cup whole cane or brown
  sugar
1 pinch baking soda (no more
  than ¼ teaspoon)
juice and rind of half a lemon
½ teaspoon ground cinnamon
2 eggs
fat for frying

POLENTA

Polenta is a traditional peasant dish made from finely ground corn. Technically, corn is a grain rather than a vegetable. Polenta requires very little preparation. A short period of pre-soaking (2 or 3 hours) in an alkaline liquid will help to release many of the important nutrients. I usually add a teaspoon of bicarbonate of soda to my bowl of polenta, then fill the bowl with water and allow it to soak for a couple of hours.

Polenta can be served wet or firm. Traditionally it was blended with a rich cheese such as parmesan or mozzarella and served as an accompaniment to a wet meat dish such as a casserole or stew. You can also allow it to set in a bread tin overnight and then toast it as you would slices of bread, or bake it with a little cooking fat and serve it alongside a main meal.

Try to use a large, deep saucepan when cooking this dish, as the polenta can spill and splatter as it cooks.

Put the polenta in a large saucepan and cover it with plenty of water. Stir through the bicarbonate of soda, then let it rest on the bench for at least 20 minutes (an hour or two is ideal).

Soaking time: 20 minutes
Cooking time: 20 minutes
Refrigeration time: overnight
Serves: 10

Ingredients:
3 cups polenta
1 teaspoon bicarbonate of soda
1 or 2 cloves garlic (optional)
1 handful arame seafood flakes
   (optional)
salt and pepper

Next, bring the water to a boil. Keep it simmering over medium heat for a good 20 minutes, partially covered. Stir it every few minutes to prevent it from sticking to the bottom.

Turn up the heat and remove the lid. Let the excess water evaporate and continue to stir every few minutes. Add the garlic and seaweed flakes (if you are using them) and season with salt and pepper to taste.

Grease a loaf tin with a little butter, then pour in the polenta. Put it in the refrigerator and let it set for 12 hours or overnight. When it sets, it can be sliced like bread and toasted or baked with a little cooking fat.

Hearty Brown Rice

Rice is currently a staple food for around half of the world's population, including much of the developing world. It is cheap to produce, but does not constitute a meal in itself; it needs to be combined with protein and vegetables to make a complete meal.

There are thought to be more than 100,000 distinct types of rice throughout the world. The most nutritious are the unrefined varieties such as brown rice and wild rice, and this hearty brown rice is a great way to give leftovers another life.

Brown rice is unmilled and comes in long-grain, short-grain, and aromatic varieties. It takes longer to cook, as its outer layers are intact and need time to break down. Just like flour, it is more susceptible to spoiling than its refined counterparts, so if you are not going to use it quickly, it is best stored in the fridge or freezer.

Put the rice, stock, and water in a medium saucepan. Bring the water to a gentle simmer and cook for 1 hour, stirring occasionally. You might need to add more water as you go.

When the rice is almost cooked, increase the temperature and boil for 3 to 5 minutes to evaporate the remaining water.

Preparation time: 2 minutes
Cooking time: 1 hour
Serves: 4 or 5

Ingredients:
1½ cups brown rice
1 cup stock
2 cups water

Alternatively, you can drain the rice through a sieve. Toss through some salt, pepper, and olive oil and serve as an accompaniment to a main meal.

Leftover roast chicken with rice:
Steam a head of broccoli and chop it into small florets. Combine it with a handful of chopped spinach and the meat left over from a roast chicken. Stir this through 1 quantity of hearty brown rice, season to taste, and add a little extra olive oil for moisture.

Leftover roast lamb with rice:
Combine 1 quantity of hearty brown rice with 1 cup of leftover roast lamb, cut into bite-sized pieces. Add 2 cups of chopped spinach, 3 finely chopped spring onions, and a handful of finely chopped mint. Season with salt, pepper, and olive oil.

GOT MILK?

"I asked the waiter, 'Is this milk fresh?' He said, "Lady, three hours ago it was grass.""

—Phyllis Diller

T O M E , G O O D M I L K E P I T O M I Z E S T H E frugavore way of eating. It's all about getting back to basics: obtaining tasty, nutritious food straight from the source and supporting sustainable farming practices. I love cows, and I love seeing what they eat when I visit my local milkman every second weekend. So it always intrigues me when people say they don't like milk or can't drink it. What sort of milk are they drinking? And when was the last time they visited a real cow in a real barn? I sometimes feel the urge to drag them out of the city, introduce them to my bovine friends, and encourage them to try a glass or two.

Milk was a staple food in many traditional diets. The dairy cow occupied a central position in traditional small-scale farms, providing a highly valued product that could be sold locally. Dairy products were thought to have unique medicinal properties. During the early twentieth century, physicians in Europe and America were known to prescribe fresh milk for a range of conditions including asthma, neuralgia, gastric disorders, and diabetes (this was sometimes called "the milk cure").

Traditionally, most dairy products were fermented. Without refrigeration, raw milk would culture in a period of days or even hours in warm weather. The naturally occurring bacteria would transform it into yogurt, kefir, cheese, and cottage cheese. Many of the famous French cheeses, including roquefort, camembert, and brie, were first created as a means to preserve milk for long periods.

Most importantly, people loved drinking milk. It was part of their culture, depicted in art and pottery, and revered as a holy food source. In countries such as India, the cow is still considered a sacred and virtuous animal.

Our attitude to dairy changed at the beginning of the twentieth century, as cities grew and milk production became more centralized. Milk was being transported over long distances without proper sanitary practices, and many people became ill from drinking it. We can now see that most of these cases involved cows being fed an improper diet of brewery waste, food scraps, and grains, while living in close confinement without access to grass or fresh air. People had little idea about sanitation or how to safely transport milk, and they didn't have stainless steel storage or refrigeration; no wonder the milk was foul. As a reaction to the many outbreaks of contamination, milk came to be pasteurized and we lost our connection to the farm and to the fresh bovine wine we once enjoyed.

Today, when I pick up a bottle of milk at my local milkbar, the product is very different from its traditional counterpart. Most milk is now reduced in fat. It is homogenized and pasteurized. That's fine from a food-safety perspective. The milk is easier to transport and has a longer shelf life. But from a taste and nutritional perspective, it leaves a lot to be desired.

Pasteurization, which gets rid of dangerous pathogens and bacteria by boiling the milk, was introduced in the early twentieth century as a food-safety measure. Homogenization became popular somewhat later, during the 1950s. It is a process whereby fat particles in the milk are broken up and dispersed throughout the liquid. The only reason for homogenization is aesthetic, but many people believe it makes the milk far blander and less palatable. Personally, I love a good layer of cream on top.

ALLERGIES & INTOLERANCE

During the past fifty years we have seen a sharp rise in milk allergies and many people, as part of their quest to live a healthier life-style, simply don't drink milk. Lactose intolerance is now one of the most common food allergies. Essentially an inability to digest the milk-sugar lactose, it is caused by the absence of the enzyme lactase in the digestive tract. This enzyme is naturally present in raw milk, but is killed off during pasteurization. Many people who are diagnosed as lactose intolerant find that they have no problem drinking their milk raw. What's more, during the traditional fermentation of products such as kefir, yogurt, and cheese, the lactase enzyme, along with the plentiful healthy bacteria that are present in fresh dairy products, breaks down the lactose in the milk even further, so that it is virtually extinguished before it is consumed.

ACCESSING FRESH MILK

Fresh milk was traditionally considered a natural superfood. Unprocessed milk is teeming with live enzymes and healthy bacteria, which aid the digestive process. It is also more nutrient-dense than its modern processed counterpart. During the pasteurization process, many key vitamins and minerals are depleted. Fresh milk production also favors small-scale, sustainable farms and heritage breeds of cow (these produce a rich and creamy milk with a high butterfat content). Today, many people are rediscovering this and looking for ways to access unprocessed milk again. They are creating "cow shares," or driving for several hours to buy their milk straight from the farm. In some places, it is not legal to sell unpasteurized dairy products. In these areas, farmers may label their wares as "bath milk" or "pet milk" in order to meet consumer demand. Sometimes there will simply be a "no transaction" system in place, whereby people collect their milk from the farm and leave some fresh food or other goods in return. I have heard of one farmer who gets his clients to pay his electricity bills while he gives them "free" milk!

In other parts of the world, an alternative system has developed to ensure healthy, top-quality milk without pasteurization. "Certified raw milk" must meet standards relating to the diet and lifestyle of the animals, sanitation, and transportation. Under this system, milk only needs to be pasteurized if it is not fit to be consumed raw. In California, for example, you can buy fresh milk, totally unprocessed and in plain glass bottles, from organic supermarkets. The largest raw-milk distributor in the United States of America, Organic Pastures Dairy Company, tests its milk for pathogens and posts the results on its website daily. They have never had any problem with microbial contamination or unhealthy milk. They attribute this to their clean pastures, healthy cows, organic farming methods, and sanitary milking conditions.

Similarly, in parts of Europe where small-scale local farming is highly valued, milk can be bought from "milk machines." These automatic milk-dispensing devices provide non-homogenized, non-pasteurized milk from cows in the local area. Above the machine is a description of the animals, the farm, and the nutritional contents of the milk. In Italy, there are now more than 250 such machines in schools, offices, hospitals, and supermarkets.

The best way to find a good source of quality milk is to tap into local food networks and find out where your milk comes from. Start by shopping at farmers' markets and getting a sense of what is available (and legal!) in your area. Websites such as www.realmilk.com give a good rundown of what to look for when you are buying milk fresh—ensure that the cows are healthy and pasture-fed, and that the correct sanitary practices are in place during processing and bottling. It is not expensive for dairy farms to test for pathogen counts, and this should be standard practice in all milk distribution.

If you're not keen to buy your milk bubbling fresh from the farm gate or local market, at least look for non-homogenized, organic milk and dairy products from grass-fed cows. This will ensure a healthier, creamier product, and you will also be supporting more sustainable farming practices. When you take your jug of milk home, be it fresh from the farm or from a local foodstore, you'll be able to make all sorts of delicious goodies, including kefir, yogurt, cream cheese, and whey. In this chapter, you'll find recipes for some of these traditional treats.

CHAMPAGNE MILK (KEFIR)

Fermented milk products such as yogurt and kefir are traditional staples in all milk-drinking societies. Food historians have documented over forty different words to describe yogurt and kefir, which shows how widely they are enjoyed. As well as preserving milk, fermentation creates healthy lactic-acid bacteria and makes other nutrients in milk easier for the human body to use.

From a nutritional perspective, traditional kefir is superior to the yogurt you find at the supermarket. It contains a diverse range of micro-organizms and wild yeasts and has a tarter, sweeter flavor. It also has a delightful fizz, hence its "champagne" nickname. It's easy to make at home; all you need is some fresh, non-homogenized milk and some kefir grains, which you can buy at health-food stores or online. Kefir grains contain the yeast and bacteria essential for fermentation. They can be re-used and will last indefinitely—so as long as you have milk, you will have an endless supply of kefir.

Combine the kefir grains and milk in a sterilized 1-quart glass jar. Place a muslin cloth or tea towel over the top and leave it in a warm, dark place. Stir or shake the glass once or twice while

Preparation time: 5 minutes
Resting time: 24 to 36 hours
Makes: about 3¼ cups

Ingredients:
¼ cup kefir grains
1 quart non-homogenized organic milk or fresh milk

Variations:
To make a smoothie, combine 1½ cups kefir with 1 or 2 egg yolks and ¾ cup fresh berries in a blender. Add a little stevia, honey, or maple syrup if you like a sweeter flavor.

it sits, making sure the grains are dispersed throughout the milk, rather than accumulating at the surface. The optimal temperature is between 66 and 84°F; depending on the temperature, the time required varies between 24 and 36 hours (the warmer the temperature, the faster the milk will ferment). Don't worry too much about getting the temperature and the timing exactly right—making kefir is all about intuition. When it is ready it should be tart-tasting and sparkly, like drinking yogurt. If it isn't, it probably needs another couple of hours' fermentation.

Once fermentation has taken place, separate the milk from the grains using a wooden spoon, or by pouring the milk through a plastic sieve. Try to avoid any contact between the kefir and stainless steel. The milk can be drunk and the grains can be re-used in the same jar with a new quart of milk.

If you've had success and are going to use your kefir jar again right away, don't wash it! If your kefir worked the first time, it means the friendly bacteria have acclimatized and will thrive with subsequent batches. If a batch hasn't worked, or if you've left a long time between batches, however, it's advisable to sterilize your jar before you next use it.

curds & whey

When you buy raw milk, you don't just get a healthier, tastier drink; you also get the ingredients you need to make other dairy products at home. Curds can be spread on bread or toast, while whey can be drunk or used as a starter in cheese-making or sauerkraut. This recipe is as old as milk, but I first encountered it in Sally Fallon's bestseller *Nourishing Traditions*, a great book about traditional foods.

Pour the milk into a sterilized glass jar. Place the jar in a warm, dark spot. As for kefir, the optimal temperature is between 66 and 84°F, and the time required will vary between 18 and 30 hours, depending on the temperature. Over time, the milk should separate into curds and whey; the whey will be a clear, yellowish liquid, while the curds are heavy white blobs.

When the curds and whey have separated, place two muslin cheese cloths (or a single cloth folded in half) over a bowl. Pour the curds and whey onto the cloth and leave it overnight; the whey will drip through to the bowl, leaving just the curds on the cloth. The whey can be drunk (it's highly nutritious) or used to make cheese or sauerkraut. The curds can be enjoyed as a spread, similar to cottage cheese.

Ingredients:
1 quart fresh raw milk

maple-syrup yogurt

Fermenting milk to produce yogurt is a traditional way to make it last longer. Natural yogurt also contains a wide range of healthy bacteria known to benefit digestion and improve the availability of the nutrients in the milk. This recipe is delicious at breakfast time, and a great way to make milk go a little further.

Put the milk into a saucepan over low heat. Allow it to warm to just above body temperature (104 to 107°F is ideal). To test it, stick a clean finger into the milk; it is warm enough when it is just warmer than lukewarm.

Rinse a large pouring jug with hot water, then pour in the lukewarm milk, yogurt, and maple syrup. Stir well. Pour the milk mixture into a sterilized 1-quart thermos and leave it in a warm spot overnight. The milk needs to stay at a constant warm temperature to turn into yogurt. If you don't have a thermos, pour it into a ceramic dish and leave it near a heater or in the oven with the oven light on.

The next morning, pour the yogurt from the thermos and enjoy with your cereal, porridge, or stewed fruit and honey.

Preparation time: 5 minutes
Resting time: overnight
Makes: 1 quart

Ingredients:
3 4/5 cups non-homogenized organic milk
½ cup natural yogurt
1 spoonful maple syrup

Tip:
To make a thicker yogurt, drain the yogurt through a muslin cloth to separate the whey from the milk solids. The whey can be re-used as a starter for sauerkraut or pineapple beer.

Lemon-curd cheese

This is a simple way to turn fresh milk into a light and fluffy lemony cheese. This is delicious spread over fruit toast, or wrapped with prosciutto and served on rye bread. Ceramic and copper cooking pots are ideal for cheese-making, but stainless steel will also do just fine.

Pour the milk and whey into a cooking pot and place over low heat on the stovetop. Stir the milk every few minutes and bring it to just above body temperature (104°F is ideal).

Meanwhile, juice the lemons. When the milk is ready, pour in the lemon juice, stirring constantly with a wooden spoon.

Let the mixture sit undisturbed for 25 to 30 minutes. The milk should separate into clumps of curds and liquid whey.

Fold a cheesecloth in half and place it over a large bowl. Pour in the separated milk, then suspend the cheesecloth over the bowl for 1 hour. After an hour, remove the curds from the cloth, add the salt, and toss them with your hands, allowing clumps to form. Return the mixture to the cloth and leave it for a further 3 to 4 hours.

Remove the cheese from the cheese cloth. It can be eaten now, or will keep in the fridge for up to 3 days.

Preparation time: 15 minutes
Resting time: 6 hours
Makes: 3 cups of cheese

Ingredients:
1 quart non-homogenized organic milk
3 lemons
½ cup whey (optional)
½ teaspoon finely ground sea salt

FATS

"Love is like butter. It is good with bread."

—Yiddish proverb

EVER SINCE HUMANS FIRST STARTED cooking, we have relied on fats to moisten our bread, enhance our vegetables, and seal our pans. Good quality, healthy fat is integral to the taste and flavor of our food. It also fulfills important nutritional requirements and imparts a wonderful feeling of satiety.

Until roughly a hundred years ago, cooks kept a variety of different animal fats in their pantry. These would have included home-rendered lard, fresh pork fat for frying, butter, poultry fat, and beef suet. Fats fulfilled a wide range of purposes and no kitchen was complete without them. The fats used in particular areas depended on which animals they raised. If a community kept cows, they used tallow and butter; if they kept sheep, they used sheep butter and lamb fat; if they were hunters, they used the fats of whatever animals they caught; if they were close to the sea, they ate marine fats from animals like seals and whales; if they kept pigs, they used lard and bacon fat. To these were added plant oils available locally—olive oil in the Mediterranean and the Middle East, sesame oil in Asia, coconut and palm oil in tropical regions.

During the past hundred years, Western diets have changed more radically than during any other time in history—and the greatest change has been the abandonment of healthy traditional fats and oils for industrial fats and oils. Along the way, we've developed a fear of fats, thanks largely to misconceptions. Fats should be a healthy staple in home-cooking, not something we are fearful of. What's more, in order to be thrifty and frugal with the animals we cook, we should be using every bit—including the fat! We need to return to the healthy habits of the past, understand where our food comes from, choose the healthiest traditional fats and avoid newfangled processed foods and transfats. Wholesome home-cooking gives a greater feeling of satiety, meaning we need only eat regular, sensible amounts to feel full.

TRADITIONAL FATS

Traditional cooking made use of every part of the beast. Nothing was ever wasted, and healthy fats were no exception. A clever housewife would never dream of discarding a precious piece of suet, tallow, or chicken fat. Today, people are rediscovering the pleasures and benefits of traditional, healthy fats—butter spread over bread, duck's drippings saved from a roast, lard put to good use as shortening in homemade pies and pastries. Organizations that promote traditional foods are celebrating the richness and flavor of traditional fats. Slow Food holds special events dedicated to offal and lard preparation, while the Weston A. Price Foundation explains the importance of the vitamins found only in animal fats, especially for growing children.

Traditional animal fats have been demonized during the past fifty years. The mainstream media have often associated their intake with cardiovascular disease. A recent analysis of all published journal articles on the topic, however, found that there was unequivocally no link between saturated fat and heart disease. Similarly, the U.S. National Academy of Sciences has found that it is not saturated fats that are to blame for heart disease, but trans-fatty acids. "Even an incremental increase in trans-fat," their report stated, "can increase the risk of coronary heart disease."

INDUSTRIAL FATS

For the first time in history, we are avoiding traditional animal fats while consuming new industrial ingredients—simple carbohydrates like sugar and white flour, hundreds of synthetic additives, and industrial seed oils (from corn, cottonseed, soybeans, and safflower seeds) in both liquid and solid (partially hydrogenated) forms. Both forms are dangerous—the liquid oils are filled with dangerous free radicals, while the partially hydrogenated oils contain equally dangerous trans-fats.

Processing methods have also changed. Traditionally, olive and sesame oils were produced by cold-pressing, whereby the oil is gently squeezed from the fruit or seed by stone presses. Modern seed oils, however, are processed using high-heat extraction, which yields a rancid, damaged oil. Solvents such as hexane are used to extract more oil, before the oil is de-gummed, deodorized and bleached to improve its appearance and flavor.

These modern ingredients entered our food supply around the turn of the twentieth century. "Oleomargarine," the first butter substitute, was invented as far back as 1869, but was initially shunned by consumers: who wanted fake butter when they could enjoy the real thing? It was only during the first half of the twentieth century that it was finally embraced, first by Americans and then by the rest of the Western world, as a cost-effective substitute for butter during war-time.

Healthy fats increase our feelings of satiety—meaning we feel contentedly full for longer. Modern seed oils, by contrast, don't fill us up, and we end up eating a lot more of them. Because they are empty of nutrients, our bodies keep telling us to eat more. As a result, we have significantly increased our caloric intake. We now eat more sugars, processed foods, refined carbohydrates, and empty, refined fats and oils than ever before, and this is why our waistlines have expanded. We are surrounded by "low fat" labels, but in fact we are eating processed foods filled with hidden fats—and the wrong kinds of fats.

Understanding fats

Fatty acids (the technical name for fat molecules) are distinguished by the presence and placement of the double bonds on the carbon chain. They can be called saturated (no double bond), mono-unsaturated (one double bond), or polyunsaturated (two or more double bonds). The chemistry of fats affects their stability under high tempera-tures and also their metabolic functioning within cell membranes. The fats we eat are always a combination of different fatty acids. We tend to classify fats, however, by their predominant fatty acid. Thus olive oil contains 14 percent saturated fats, 71 percent mono-unsaturated fats, and 15 percent poly-unsaturated fats, so is classified as largely mono-unsaturated. Traditional animal fats contain mostly saturated and mono-unsaturated fatty acids.

In contrast to what you may have heard, the safest fats are saturated and mono-unsaturated fats. Saturated animal fats are the best fats to cook with, as they are stable at high temperatures and don't oxidize under high heat. Mono-unsaturated fats are less heat stable, while polyunsaturated fats are *highly* sensitive to heat and can oxidize easily without refrigeration. Mono-unsaturated olive oils and sesame oils have traditionally been used for cooking but are in fact best used unheated, for instance in salad dressings.

Trans-fats

Traditional seed oils such as soy, canola, and cottonseed oils are liquid at room temperature, which makes them difficult to cook or bake with. The food industry gets around this problem by partially hydrogenating seed oils to produce fats with a higher melting point and longer shelf life. These new "trans-fats"—which are also very cheap—have displaced natural animal fats in many areas, notably in the fast-food, snack-food, fried-food, and baked-goods industries.

Traditionally, polyunsaturated fatty acids were consumed in small amounts, as traces of them are found naturally in most foods. Today, however, the average person gets about 30 percent of his or her calories from vegetable oils, both liquid and partially hydrogenated—much to the detriment of our health. Trans-fats accumulate in our cell membranes, where they inhibit many chemical reactions. They increase our risk of developing cardiovascular disease, interfere with insulin production, and can contribute to obesity and diabetes.

Olive oil

Olives are fruits, not seeds, so technically olive oil is not a seed oil. It was traditionally extracted using cold-pressing and most producers still use this method. Look for cold-pressed, extra virgin oil, which indicates the healthiest extraction method. Olive oil is rich in antioxidants and in mono-unsaturated fatty acid, a healthy fat known to benefit heart health.

*

The animal fats eaten by our grandparents, and by the slender French, came from free-ranging, grass-eating animals. These animals were not kept in feedlots and fed piles of grains, soy, or cornmeal, nor were they subjected to overcrowding or antibiotics. Grass-fed fats are much cleaner and safer, and also higher in key vitamins like A, D, K2, and E.

They contain CLA, a type of fat that prevents cancer and weight gain. Choosing healthy fats is not just about choosing butter over margarine—it's also about looking for locally raised, free-ranging animals raised on a natural diet with plenty of sunlight and exercise.

Healthy fats are not only integral to our body's natural metabolism—they are also essential to any home kitchen. Learn to be thrifty with your food—render your duck fat when you are cooking a roast, save your delicious bacon fat, savor some lard from your local butcher, discover the stable qualities of tallow for frying. As well as being better for you, traditional fats are incredibly frugal, as they are made using parts of the animal you might otherwise throw away. In this chapter, you'll find a collection of recipes for old-style fats including home-rendered lard, goose fat, and ghee, just like your grandmother probably cooked with. I promise they'll add a whole new dimension of flavor to your cooking.

A note on the nutritional profile tables: The fatty-acid profile can vary with the breed of the animal and its lifestyle and diet. The numbers do not always total 100, as there is also connective tissue and water in the fat.

PORK FAT

Nutritional profile: Saturated fat 39 percent; mono-unsaturated 45 percent; polyunsaturated 11 percent

Pork fat, or lard, is a very economical fat for home-cooking. It can be rendered from several different parts of the pig, each of which produces its own unique flavor. Fat from the back, shoulder, or rump of the animal, all described as "back fat," is the most abundantly available and produces a good, hearty lard. "Leaf lard," the fat from around the pig's kidneys, is ideal for baking and is the most prized of all pig fats.

To make lard at home, preheat your oven

to 230°F. Roughly chop the pork fat into 1-inch pieces or as small as possible. Place in a heavy baking dish and cover with 1 inch of water. Cook in the preheated oven for 4 to 12 hours. Keep an eye on the water levels and top up if necessary; there should always be at least ¼ inch of water in the dish. The lard is ready when the chunks of fat have become thin and crackly looking.

Remove the dish from the oven and strain the contents through a muslin cloth into a large bowl. Transfer the lard to glass jars and store it in the fridge until you are ready to use it.

POULTRY FAT

Nutritional profile:

CHICKEN: Saturated fat 30 percent; mono-unsaturated 45 percent; polyunsaturated 21 percent

DUCK: Saturated fat 33 percent; mono-unsaturated 50 percent; polyunsaturated 13 percent

TURKEY: Saturated fat 29 percent; mono-unsaturated 43 percent; polyunsaturated 23 percent

Poultry fat is the favored cooking fat in the west of France. A recent study found that people in regions that consume large amounts of poultry fat have the lowest incidence of cardiovascular disease. Poultry fat adds a delicious flavor to cakes, hot potatoes and general cooking.

To render fat from a fresh bird, trim the fat and excess skin from the carcass. Cut them into small pieces, place them in a small saucepan, and cover them with water. Bring to a boil, then reduce the heat and simmer for 45 minutes or until the fat is completely rendered and the skin and connective tissue is crisp (this "crackling" is delicious on salads). Strain the fat into glass jars and refrigerate.

To render fat from a roasted bird, heat the drippings from the roasting pan until they are warm and runny. Sieve them through a muslin cloth (for cleaner-looking fat), or simply pour them straight into a clear glass jar. Store in the refrigerator.

BUTTER

Nutritional profile: Saturated fat 65 percent; mono-unsaturated 30 percent; polyunsaturated 4 percent

Ingredients:
2 cups cream, at room temperature
iced water as required
Optional: sea salt, paprika, or chili to taste

Place the cream in a small electric blender or food processor and pulse on the highest setting. Very gradually pour in the water, a few drops at a time. After a few minutes you should notice particles of butter appearing. These will get bigger and bigger until they form large yellow clumps.

Drain the liquid from the solids. The liquid is buttermilk, and the solids are butter. Place the butter in a bowl and press with a wooden spoon to work out the liquid. Rinse the butter with water and repeat the process several times—butter needs to be well washed or it will develop a rancid smell.

Variation: You can add a little sea salt or even some spices (such as paprika or chili) before you start churning the cream. If you use cultured cream or cream from kefir-making you will get a delicious, sweeter butter.

GHEE

In India, to preserve butter for longer periods of time, milk solids were heated to make ghee. Ghee is very stable, so it is perfect for high-temperature cooking such as frying and baking.

To make ghee, place one cup of butter in a small saucepan. Slowly bring it to a gentle simmer, then let it simmer until the milk solids begin to separate. The milk solids will be frothy and brown, while the rest will be a rich golden color. Skim off the solids that rise to the top and drain the golden liquid into glass jars, leaving any solids that have sunk to the bottom: this is your ghee. The solids can be discarded or, as in India, used in cake and toffee-making.

Suet

Nutritional profile: Saturated fat 75 percent; mono-unsaturated 22 percent; polyunsaturated 2 percent

Suet is the fat that surrounds an animal's kidneys, and is usually made from beef or lamb. Because it is so saturated, it has a high melting point, which makes it excellent for deep frying, pastry-making, and puddings.

Preheat your oven to 250°F. Chop the fat into small pieces, place it in an ovenproof dish, cover it with water, and cook in the preheated oven for 4 to 8 hours. Alternatively, gently simmer it on the stove top on low heat, stirring occasionally, for the same period of time.

Tropical Oils

Nutritional profile:

Coconut oil: Saturated fat 92 percent; mono-unsaturated 6 percent; polyunsaturated 2 percent

Palm oil: Saturated fat 40 percent; mono-unsaturated 41 percent; polyunsaturated 9 percent

Palm oil and coconut oil are tropical oils, traditionally used in areas where these plants natively grow. Coconut oil is delicious to cook with; its high saturation makes it perfect for high-temperature cooking, baking, and frying. Palm oil has a mildly disagreeable flavor and is often a lot more expensive than coconut oil. Always look for cold-pressed, extra-virgin varieties of tropical oils, and make sure that they have been sourced sustainably.

THE SWEET STUFF

"Life is uncertain. Eat dessert first."

—Ernestine Ulmer

THE WORD *DESSERT* IS TAKEN FROM THE French verb *desservir*, meaning to clear the table following a meal. As this suggests, the traditional purpose of dessert is to refresh and nourish guests after eating, not to fill them up with refined sweeteners and processed food. So in keeping with the frugavore ethos, I've tried to make these desserts as simple and nutritious as possible. The cakes are not the usual fluffy afternoon-tea variety. They are more like the English pound cakes our grandmothers used to make: denser, less sweet, but very satisfying and much better for you.

As humans, we naturally crave a sweet flavor in our food. Since the beginning of time, we have gone to great lengths to obtain natural sweeteners such as fruit, honey, molasses, or maple syrup. This might have involved braving a swarm of bees to get a few teaspoons of honey, or tapping a maple tree for a drizzle of syrup. Natural, traditional sweeteners have many nutritional benefits that are not found in their modern, refined counterparts. For instance, blackstrap molasses is sourced directly from sugar cane. It contains iron, B vitamins, calcium, and trace minerals and is known to have many health-promoting properties. Modern table sugar is also sourced from the sugar cane plant—but thanks to intense processing and refinement, it is devoid of nutrients and is implicated in modern health problems such as diabetes, obesity, and nutrient deficiencies.

Not surprisingly, as new sweeteners have become available, our dessert recipes have had to adapt. Back in 1400 AD, gingerbread was made by soaking breadcrumbs in a mixture of honey and spices. Chocolate was traditionally a spicy drink made with bitter-tasting cocoa beans. Nomadic wanderers through the desert would wrap dried figs in animal fat for a daily sugar fix. Eskimo cultures would whip together berries with seal fat to make ice cream. Today, we have pure sugar lollipops, richly flavored chocolates, and soft drinks aplenty. Our sweets are overly refined and stripped of most of their nutrients.

There are alternatives, however. Most health-food stores now stock a wide variety of natural sweeteners, which can be used in place of conventional sugar in most recipes. Alternatively, opt for old-fashioned sweeteners such as ripe fruit, fresh dates, or a few drops of honey or molasses.

A glossary of sweeteners

Many people are confused about which sweeteners to use for home-cooking. Here's a run-down of some good and bad options.

Agave syrup: Made from the sap of various species of agave desert plants, which are related to the cactus family. Just because it is "plant-based," however, doesn't mean it's any better for you than standard sugar. Many brands of agave syrup are highly processed and have a similar nutritional profile to conventional white sugar.

Brown sugar: The first industrially produced

brown sugars were by-products of turning cane juice into unrefined sugars. These included demerara, turbinado, and muscovado. They had some nutritional value and were vastly superior to modern refined sugars. Nowadays, brown sugars are produced at the refinery using raw sugar. Molasses is added to the sugar to impart a more complex flavor.

From a nutritional perspective, brown sugar is only marginally better than white sugar. But I have to admit, I still use it when I am making sweets in bulk—e.g., jam or a large batch of cakes—as it is extremely cost-effective. Just try to consume these foods in moderation.

Chocolate: The chocolate we buy from our local candy store is rich in sugar and preservatives. For cooking purposes, look for dark chocolate (at least 70 percent cocoa) and free of preservatives (including soy lecithin).

Dates: Dates can be puréed in a food processor and used as a substitute for sugar—one cup of dates can replace a cup of sugar. They are a good alternative in many baking recipes.

Honey: The nectar of the delicate honey bee. There is nothing quite as special as honey. Look for raw honey wherever possible (many of the good nutrients are killed off during pasteurization). Many inner-city foodies are now starting their own hives, which can be kept on rooftops or apartment balconies in the inner city. Honey contains unique antibacterial properties and has been used as a traditional remedy by many cultures.

Lakanto: Although not widely available in the West, Lakanto is a natural sweetener derived from the *luo han guo* fruit of China. Lakanto has no kilojoules and does not raise blood-glucose levels at all. It can be used in baking as a substitute for sugar.

Malt syrup: An ancient and versatile sweetener. Along with honey, this was the primary sweetener in China for 2,000 years. It is made from a combination of germinated cereal grains, especially barley, and ordinary cooked grains. Malt syrup is considerably less sweet than sugar syrup, but can be used in baking.

Maple syrup: The sap of the maple tree, boiled until it forms a smooth syrup. It is delicious on pancakes and porridge and can be used in place of normal sugar for baking purposes.

Molasses: Also known as treacle, this is the syrup left over after cane-sugar processing. Blackstrap molasses is the most nutritious of all the molasses varieties produced during sugar processing.

Whole cane sugar or sucanat: This is dehydrated cane juice, and is a less refined version of conventional white sugar. Whole cane does raise blood sugars, but not nearly as quickly or intensely as conventional sugar. This is a useful alternative to sugar and can be used for baking. It can be expensive, so try to buy it in bulk or at a wholesale outlet.

Stevia: This is a natural herb, native to South America. Stevia can be grown as a plant in your own backyard: I have several pots sprouting around my kitchen door. Many health-food stores sell it in powdered or liquid form. These are extremely, exquisitely sweet: one drop equals one cup of sugar, but does not raise blood-glucose levels at all, so has none of the negative effects associated with sugar. It is useful as a sweetener in liquid drinks, but I do not recommend it for cake-making.

White sugar: Of all the sweeteners mentioned, this is the sweetener that I would go to great lengths to avoid. Because white sugar is highly refined, it is rich in kilojoules but contains no nutrients whatsoever.

MINTY LEMONADE

This is a delightfully refreshing drink, sweet and tangy at the same time. It's a great way to use up all that mint spilling from your backyard pots. Stevia does not raise blood sugars at all, so it is an excellent alternative to conventional sugar, and this drink is a healthy alternative to conventional lemonade.

Thinly slice the lemons and finely chop the mint. Add them and the stevia to your jug of mineral water and stir well. Add some ice and it's ready to drink!

Preparation time: 5 minutes
Serves: 4

Ingredients:
2 medium lemons
1 large handful fresh mint
3 leaves stevia, crushed, or
   2 drops stevia liquid
1 quart sparkling mineral water

HOT APPLE CIDER

Bev Smith first introduced me to this recipe when I ventured out to visit her farm in Gippsland during the middle of winter. It's a wonderful combination of honey, apple cider, and nourishing warmth. I recommend it first thing in the morning before breakfast, or last thing at night before you sleep.

Place ½ teaspoon apple-cider vinegar and 1 teaspoon of honey in each mug. Fill each mug with cold water so that it is half full. Boil a kettle of water and fill both mugs. Stir well and drink immediately.

Preparation time: 5 minutes
Serves: 2

Ingredients:
1 teaspoon apple-cider vinegar
2 teaspoons honey
2 cups water

BREAD & BUTTER PUDDING

When you realize late on a Sunday night that you have a hungry family roving your kitchen and nothing to feed them for dessert, this is the ultimate leftover dish, quick to prepare with a few simple ingredients. Any bread will do—I've made it with sourdough hot cross buns at Easter, and leftover panettone after Christmas. Here is the basic recipe, but you can also add fresh fruit, some jam from your last harvest, or whatever happens to be in season.

Preheat your oven to 350°F.

Butter the sides and bottom of a small or medium baking dish about 2 inches deep. Generously butter one side of each slice of bread and neatly pack the buttered bread into the dish.

In a small bowl, whisk together the sugar, eggs, milk, cinnamon, and vanilla. Pour this mixture over the bread so that the bread is three-quarters immersed, then leave the pudding to settle and soak for 30 minutes at room temperature. Sprinkle the currants between the layers of bread and bake in the preheated oven for 40 minutes, or until the liquid is set and the bread is lightly toasted.

Preparation time: 10 minutes
Soaking time: 30 minutes
Cooking time: 45 minutes
Serves: 6

Ingredients:
butter
7 slices bread
¾ cup whole cane or brown
  sugar
4 eggs
¾ cup milk
1 pinch cinnamon
½ teaspoon vanilla essence
⅓ cup currants

Variations:
Replace the currants with 1 cup of fresh berries or thinly sliced nectarines, peaches, or plums. Or, add a few dollops of jam to each slice of bread. Apricot jam works well.

Ginger Cake by Giules

Ginger smells exquisite when baked—the aroma permeates throughout the house, evoking wonderful thoughts of ginger-bread and cookies. My friend Giules created this cake when there was little fresh food in the kitchen save some fresh ginger root and a lemon from our back doorstep. Despite its frugality, this is a deliciously moist cake. It goes wonderfully with fresh cream and stewed fruit.

Line a 8-inch cake tin with greaseproof paper.

Combine the flour with the yogurt and milk to make a smooth paste. Cover the bowl with a tea towel and leave it to sit at room temperature. Twenty-four hours is ideal, but if you're in a hurry an hour or two is enough.

Use a food processor's gentle setting to combine the coconut fat, eggs, sugar, and honey. Add the remaining ingredients one by one. Lastly, add the flour paste.

Pour the mixture into the lined cake tin. Bake for 1 hour at 350°F. When the cake is done, the inside should be very moist but not runny, and the outside edge should be crisp.

Preparation time: 10 minutes
Soaking time: at least 2 hours
Cooking time: 1 hour
Serves: 5

Ingredients:
1½ cups spelt flour
1 cup plain yogurt
½ cup fresh milk or coconut milk
⅖ cup coconut fat or other cooking fat
4 eggs
¾ cup brown sugar or whole cane sugar
¼ cup honey
4 teaspoons freshly grated ginger
juice of 1 lemon
1 teaspoon dried cinnamon
¼ teaspoon baking soda

COCONUT SAGO

Sago is also known as "seed tapioca"—tiny balls of starch extracted from the trunk of the sago palm in countries including New Guinea and the Moluccas. The little balls are the size of caviar, and kids will often refer to them as "fish eggs" when they are cooked. Sago combines nicely with coconut milk, and this warming recipe is delicious served with fresh or stewed fruit and some fresh cream.

Place the sago in a saucepan, pour over the coconut milk, and leave it to soak. The longer it soaks, the quicker it will cook: a couple of hours is ideal, but 30 minutes is plenty.

Add the milk and place the saucepan over low heat and gently cook for 30 to 45 minutes. Sago has a tendency to stick to the saucepan, so keep a watchful eye on it. When the sago is ready, the tiny balls will be clear and soft, not hard or crunchy. If it seems too firm, you may need to add a little more milk as it cooks.

When the sago is done, add the vanilla, sugar, egg yolks, lemon juice, and lemon rind, and stir until they are well dissolved. Remove from the heat and serve with fruit or cream.

Preparation time: 10 minutes
Soaking time: at least 30 minutes
Cooking time: 30 minutes
Serves: 5

Ingredients:
1 cup sago
2 cans (15¼ ounces each) coconut milk
½ cup milk, plus a little extra
1 teaspoon vanilla essence
⅓ cup whole cane or brown sugar
2 egg yolks
juice and rind of 1 lemon
cream and fresh or stewed fruit (optional, to serve)

APPLE & NECTARINE SHORTCRUST TART

Use the freshest fruit you can find for this recipe. If you have a fresh batch of homemade jam, it will be even more delicious.

Preheat your oven to 350°F.

Grease a 8-inch tart tin with a little cooking fat. Press the pastry into the tin using your fingers (you won't need a rolling pin). Push it up the sides so that it is ¼ or ¾ inch high. Place the tin in the oven and bake for 5 minutes or until lightly crisped.

Prepare the filling by combining the cream, vanilla, sugar, egg yolks, and almond meal in a small mixing bowl and stirring well. When the pastry is ready, pour the filling in.

Wash and core the apple and slice it as thinly as possible. Arrange the slices on the pastry base in a circular pattern. Do the same with the nectarine, layering it on top of the apple but letting slivers of apple show through.

Preparation time: 20 minutes
Cooking time: 40 minutes
Serves: 4

Ingredients:
1 quantity sweet oatmeal pastry
    (see recipe on page 252)
1½ tablespoons cream
3 drops vanilla essence
5 teaspoons whole cane or
    brown sugar
2 egg yolks
⅔ cup almond meal
1 green apple
1 nectarine or plum
1½ tablespoons apricot or
    plum jam

Use a small sieve or tea strainer to dribble droplets of jam over the fruit. If the jam is too thick, you may need to combine it with ¾ tablespoon of boiling water first.

Place the tart in the oven and bake for an additional 30 minutes or until the fruit and filling are fully cooked and lightly browned. Serve with cream.

Baked Fruits Stuffed with Ricotta & Honey

This is a wonderful summer dessert. Choose fruit that is well and truly ripe; if you are picking it yourself, it should be almost ready to fall from the tree. You can make the ricotta mixture in advance and simply combine it with the fruit just before you start dinner.

Preheat your oven to 350°F.

In a small mixing bowl combine the ricotta, lemon, and sugar and stir well. Whisk in the egg to form a smooth paste.

Cut each piece of fruit in half using a sharp knife. Remove the pits or seeds from each piece and lay them hollow-side up on a baking tray.

Spoon a dollop of the ricotta mixture into the hollow of each piece of fruit and balance them carefully on the baking tray.

Bake in the preheated oven for 30 minutes or until the ricotta is lightly browned and the fruit is well cooked and soft.

Preparation time: 10 minutes
Cooking time: 40 minutes
Serves: 10

Ingredients:
⅘ cup ricotta cheese
juice and finely grated rind of
　1 lemon
¾ tablespoon brown sugar or
　whole cane sugar
1 egg
½ teaspoon dried cinnamon
10 medium or 20 small fruits
　(nectarines, apricots,
　peaches, or pears)

CHOCOLATE MOUSSE

What could be simpler and more enticing than a delicious chocolate mousse with only two ingredients? This is hands-down the easiest and most delicious chocolate mousse recipe you will ever try.

Separate the eggs, putting the yolks and whites into separate bowls.

Break the chocolate into small pieces and place them in a small ceramic mixing bowl. Put the bowl in a saucepan with an inch of water and place the saucepan over medium heat. Increase the temperature until the water simmers but does not boil. Stir the chocolate until it melts to a smooth liquid.

Whisk the egg whites until soft peaks form. When they're ready, you should be able to turn the bowl upside down without them sliding out.

Combine the egg yolks with the melted chocolate. Gradually add this mixture to the egg whites, stirring gently with a whisk until well combined.

Pour the finished mixture into individual soufflé bowls or small glasses and refrigerate for at least 2 hours. Serve with fresh fruit or berries.

Preparation time: 20 minutes
Refrigeration time: at least 2 hours
Serves: 10

Ingredients:
8 eggs
7 ounces organic dark chocolate (55 to 70 percent cocoa)
fresh fruit or summer berries (to serve)

baked custard with rum

This cake is similar to the traditional "Far Breton" from Brittany in France. Every time I cook it I am reminded of the humble but delicate cooking methods that still exist in that region. You can bake this as a single cake, or divide the mixture into soufflé dishes and serve it as individual custards. The combination of prunes, rum, and custard is exquisite.

Place the flour, sugar, milk, eggs, and yogurt in a mixing bowl and stir to form a smooth batter. Place this mixture in the fridge and allow it to chill for 5 to 8 hours (the longer the better).

Place the prunes in a small bowl and cover them with the rum. Leave them to soak for a similar period.

When you are ready to make the cake, preheat your oven to 350°F. Combine the prunes with the milk mixture and stir well.

Grease a single cake tin or 8 individual soufflé dishes with butter, or line them with greaseproof paper. Pour in the custard mixture, making sure the prunes are evenly distributed throughout the dish.

Preparation time: 10 minutes
Soaking time: 8 to 10 hours
Cooking time: 45 minutes
Serves: 6

Ingredients:
½ cup flour
½ cup whole cane or brown
    sugar
2 cups milk
4 eggs
⁴/₅ cup plain yogurt
1½ cups pitted prunes
¼ cup rum
butter for greasing

Bake for 30 minutes if you are using soufflé dishes or 50 minutes if you are using a cake tin. The cake will be ready when the custard is lightly browned and a skewer comes out clean. Serve immediately.

STEWED PEARS WITH CINNAMON SYRUP

My mother used to whip this up with whatever fruit she happened to have on hand, a little wine, and a fresh stick of cinnamon. She always served it with a hearty dollop of cream or yogurt. The key is not to smother the pears with liquid—add the minimum amount of juice and tightly pack all the ingredients into a small, heavy saucepan with a tight lid. Ideally, the pears should fit snugly and take up three quarters of the space in the pot.

Rinse the pears and cut them in half. Using a small, sharp knife, remove the seeds and the stems.

Place the pears in a small saucepan and add the sugar, vanilla, and wine. Break the cinnamon stick in half and add it to the saucepan. Peel a long strip of rind from the lemon and throw this in too.

Add enough water to the saucepan to just cover the pears—don't add too much, or the sauce will be watery. Firmly secure the lid and bring the liquid to a gentle simmer. Cook for 45 to 60 minutes on low heat, stirring occasionally.

When the pears are soft and light brown, add the berries (if you're using them). Juice the lemon and add the juice to the pot.

Preparation time: 10 minutes
Cooking time: 1 hour
Serves: 6

Ingredients:
3 large green or yellow pears
½ cup whole cane or brown sugar
a few drops vanilla essence
¼ cup dry white wine
1 large stick cinnamon
1 lemon
1 cup fresh strawberries or raspberries (optional)

Check the water level; if the liquid doesn't cover three-quarters of the fruit, you might want to add a little extra. Remove the lid, increase the heat, and simmer for 5 to 10 minutes so that the sauce thickens and the berries soften.

Transfer the pears to individual serving bowls. Pour on some syrup and drop a dollop of cream or yogurt into the hollow of each piece of fruit.

Note:

You can make this dish sugar-free by replacing the sugar with 1 cup of dried dates. Purée the dates in a food processor until they form a smooth paste, and use this in place of the sugar.

peachy mint salad

This is a very simple dessert that never fails to please. Grab the ripest fruit from your or a neighbor's backyard. Add some maple syrup and a handful of finely chopped mint, squeeze on some lemon juice, and *voilà*! Serve with a little cream, some plain yogurt, or even coconut-flavored sago. A little dessert wine doesn't take away anything, either.

Thinly slice the nectarines or peaches and arrange them in a large serving bowl.

Combine the syrup and lemon juice in a small jug. Pour the mixture over the fruit, add the mint, and stir well. Serve with fresh cream or yogurt.

Preparation time: 10 minutes
Serves: 8

Ingredients:
6 nectarines or peaches
¾ tablespoon maple syrup
juice of 1 lemon
1 handful fresh mint, finely
    chopped
cream or yogurt (optional,
    to serve)

PRESERVES FOR THE PANTRY

"Food preservation techniques can be divided into two categories: the modern scientific methods that remove the life from food, and the natural 'poetic' methods that maintain or enhance the life in food."

—Eliot Coleman

THE TRADITIONAL PEASANT PANTRY contained a wide range of preserves made from seasonal foods that were only available fresh at certain times of the year. Sheer necessity dictated that raw ingredients be stored for later use. We may not face the same shortages today, but preserves can still be incredibly cost-effective. Ever wonder what to do when your neighbor's lemon tree is overflowing into your back garden? Or when a pile of onions goes on sale at your local supermarket? Buy up big, I say, and preserve them for the future. Most preservation methods require only basic ingredients such as salt or vinegar. Other flavorings, like herbs and spices, can be added if you wish.

Preserves also make wonderful presents. Organic food stores sell stylishly packaged preserves, replete with frilly ribbons and raffia, for twenty dollars or more per jar. When you see how quick and inexpensive they are to make at home, you'll realize how ridiculous this is!

Before you start preserving, you need to remember one important thing: never throw anything out. Old jam jars, glass bottles, glass food containers—keep, keep, keep. When you are pounding your cabbage or bottling your lemons, you will make good use of them, and will be glad you kept them in your cupboards for all those years.

Preparing your kitchen

Choosing containers: Always try to use glass if possible. In some instances, ceramic cookware can be used, but make sure that it is properly sterilized. Plastic should be avoided wherever possible.

Steriliszation: To sterilize glass jars and cooking equipment (excluding plastic), first place your containers in a large pot of cold or warm water. Make sure they are fully immersed. Put the pot over a gentle heat and bring the water to a boil. Reduce the heat and simmer for five minutes. Remove the containers from the water and allow them to cool. Please note: if you add glass jars to a pot of already boiling water, they will most likely crack!

Hygiene: Many of the most dangerous microbes are transferred via our hands, skin, and hair. Follow these golden rules to avoid any contamination. Many of the recipes in this chapter require a lot of hands-on stirring and mixing. With clean hands, you can feel free to handle food as much as you like.

- Wash your hands thoroughly with soap and water before getting started, and after any breaks or trips to the toilet.
- Keep your fingernails short. This is super important, as bacteria can sit under your nails and come out as you cook.
- Before a major cooking session, soak your hands in a bowl of warm water and vinegar for five minutes.
- Keep your hair off your face and avoid any nose blowing.

Preserving with Sugar

Sugar reduces the water content of foods and prevents bacteria from growing. Obviously sugar is not in any way nutritious and should only be consumed in moderation. I try to limit my sugar intake to the odd batch of jam and a little quince jelly here and there. It's a shame to waste beautiful fresh fruit when it is in local abundance, and jam-making is a traditional way to keep fruit available throughout the seasons.

If you're trying to avoid plain sugar, you can use whole cane sugar in these recipes. Whole cane sugar is better for you, but is more expensive to buy and also doesn't seem (in my experience) to allow the jam to crystallize properly. I prefer to use raw sugar and just enjoy my jam in moderation.

Preserving with Salt

Sea salt is a natural preservative that reduces the water content of produce, prohibiting the growth of micro-organizms. Look for sea salt or Celtic sea salt for the best results.

Fermented Foods

The traditional pantry included a wide array of fermented foods: sourdough bread, homemade beer, wine, sauerkraut, pickled beets, garlic, and carrots. Fermentation is a special method of food preservation whereby microbes create a rich array of healthy lactic-acid bacteria, which make foods last longer. Fermented foods are what nutritionists sometimes call "functional foods"—that is, they have added nutritional value. Fermentation increases the nutrient-density of the raw ingredients.

Before refrigeration was invented, foods had to be fermented, preserved with vinegar, air-dried, salted—or eaten pretty darn quickly. Fermentation was an easy and inexpensive method of preserving food used by many traditional cultures. Dr. Weston A. Price recorded Eskimo cultures allowing their fish to "rot" for months before consuming them. Fermented cabbage in the form of sauerkraut, choucroute, or kimchi is a mainstay of Eastern European and Asian cuisines. In Australia, Aboriginal groups buried yams in the ground for several months, allowing them to ferment before eating them. Traditional fermentation was done simply, with just the basics—a container, some salt, a few herbs, and any vegetables not wanted for cooking. Often they were left for months and months, either in a cellar or buried in the ground.

Sadly, with the advent of modern refrigeration, along with changes in our lifestyle and climate, these fermentation practices have been largely lost or forgotten. Pollution, a dryer climate, and increased pesticide use mean that the growth of good bacteria can sometimes be inhibited. For this reason, when trying traditional fermentation techniques at home, it's important to use the best quality ingredients you can, and to make sure everything is properly sterilized.

People often ask me if they can use vinegar to ferment vegetables. In supermarkets you can buy sauerkraut made with vinegar and salt. But cabbage cooked with vinegar is not real sauerkraut. Vinegar works well as a preserving agent, but it does so by *inhibiting* the growth of micro-organizms and bacteria; it stops harmful micro-organizms from growing, but it also prevents true fermentation from taking place. The recipes here therefore don't include any vinegar.

The dos & don'ts of fermentation

- Always use organic, biodynamic, or home-grown produce. Commercial sprays and pesticides have anti-fungal and anti-insect properties, which inhibit fermentation.
- The vegetables *must* be completely submerged in the liquid. Problems can occur if they are left exposed to the air.
- You must use non-chlorinated water. Buy some filtered water, or simply boil your

own tap water for ten minutes, then allow it to cool before you use it.

- Use good quality sea salt, which is rich in a wide variety of trace minerals. Don't use table salt, which is highly refined, contains few trace minerals, and contains additives.

Starter cultures

Starter cultures are used to kick start fermentation. Adding a starter is optional, as most traditional fermentation will happen naturally without one. However, I recommend using a starter, as it ensures a foolproof, well-fermented final product. You can use the fermented whey from cream cheese and whey, a commercially made starter culture, or, if you are making beer, you can make your own ginger starter using the recipe on page 320.

poor-man's orange marmalade

My mom recently returned from South Australia and managed to stuff 4⅖ pounds of very special "poor-man's oranges" into her suitcase to bring home. In an effort not to waste them, we preserved them as marmalade that very evening. Seville oranges can also be used in this recipe and are just as delicious (but "poor man's" has a certain ring to it). Whatever oranges you use, a jar of homemade marmalade makes an excellent Christmas gift.

Citrus seeds are a blessing in jam-making—they supply a natural source of pectin, which allows the jam to set to the desired consistency. Maggie Beer suggests separating the seeds, placing them in a muslin bag and cooking them with the jam. I personally prefer the added character of seeds running loose throughout the marmalade (the odd crunchy bit doesn't seem to bother me), but you can do the muslin-bag option if you prefer. I've enjoyed this marmalade on sourdough bread with lashings of butter. It's also delicious with a soft cheese such as chèvre or camembert.

Preparation time: 20 minutes
Soaking time: overnight
Cooking time: 1 hour
Makes: 10 jars

Ingredients:
4⅖ pounds oranges
6 cups sugar
juice of 1 lemon

Thinly slice the oranges and place them in a large saucepan (a copper or enamel pot works best, but stainless steel is just fine). Be sure to include the seeds. Cover the oranges with water and leave them to soak for 12 hours or overnight.

Transfer the pot to the stove and turn on the heat. Simmer until the rind is just tender (usually about 15 or 20 minutes). Add the sugar and lemon juice and bring to a rapid boil. Cook until the marmalade reaches its "setting point," usually 40 to 50 minutes. Test it by dropping a spoonful onto a small saucer and putting it in the refrigerator for 5 minutes. After 5 minutes, it should be firm and set like jelly.

Let the marmalade cool a little, then pour it into sterilized glass jars and seal them with tight-fitting lids. Because this recipe contains less sugar than conventional marmalades, be sure to store it in a cool, dry place to prevent mold from forming on the surface.

preserved caramelized onion

This is a sweet and delicious condiment that complements any meal. I *love* caramelized onion with dry biscuits and a rich, runny cheese, with sausages and sauce, with marrow on toast, or with crispy fried liver. It can also be added to casseroles and stews for a sweet tangy flavor, and is delicious on sandwiches. The natural sugar in the molasses works as the preserving agent, while the slow cooking means the onions lose most of their water content, giving rise to an intense flavor.

Coarsely peel and finely chop the onions. Don't worry if they are not perfect—save your tears and get them into the pot as quickly as possible. Put them and the butter into a large, heavy saucepan and place over a low heat. You may want to use a heat diffuser to ensure a long, slow, even cook.

Add the molasses and leave the onions to cook over a low heat. Stir them with a wooden spoon every half hour or so. They should cook for about 5–6 hours, but keep a close eye on them. They shouldn't brown or burn; slow, gentle cooking is crucial.

Preparation time: 10 minutes

Cooking time: 5–6 hours
Makes: 2 or 3 jars

Ingredients:
3⅓ pounds brown onions
1 teaspoon butter or fat
2¼ tablespoons molasses

When the onions seem nearly done, increase the heat for about 10 minutes to ensure that they reach the perfect consistency. They should be lightly browned and have lost all their water. When they are ready, add the molasses, stir well, and remove the pot from the heat.

Allow the mixture to cool, then spoon it into sterilized glass jars. As you fill the jars, pack the onions down so that there are no air bubbles. The onions should last for several weeks if you keep them in a cool place such as a cellar or refrigerator.

QUINCE JELLY

Quinces make exquisite jelly. It's delectable with a strong cheese or on toast with a little butter.

Give your quinces a good scrub and remove the stems and any excess fluff. Cut them lengthwise into quarters, keeping the cores and the seeds intact. Place the slices in a large, heavy-based cooking pot.

Cut a long, thin slice of peel from the lemon and add it to the pot along with 3 quarts of water. Turn on the heat and bring the liquid to a boil. Cover the pot and leave it to simmer for 2½ hours.

Remove the quinces from the pot with a slotted spoon. Place them in a muslin cloth and allow them to drip into a bowl overnight.

In the morning, measure the liquid from the bowl and then pour it back into the cooking pot. For every cup of liquid, add ½ cup of sugar. Add the lemon juice and seeds, then bring the liquid to a boil and simmer for a good half hour, so that the jelly thickens and any excess water evaporates.

Preparation time: 20 minutes
Cooking time: 12 hours
Makes: 2 or 3 jars

Ingredients:
6 medium quinces
juice and seeds of 1 large
    lemon
5¼ cups sugar

To test if the quince jelly is ready, place a spoonful on a plate and put it in the fridge for 5 minutes. After 5 minutes, it should be set like jelly and not pour off the plate. When it's done, pour it into sterilized glass jars and seal them with tight-fitting lids.

preserved lemons

Preserved lemons are one of the easiest things to make in your home kitchen, and they are delicious added to roast chicken, lamb dishes, or sliced into thin slivers and added to salads or desserts for an intense lemony flavor. Don't worry too much about measuring the lemons or salt; it's easier just to add them in proportion to what you've got on hand. Many things can be added to the basic recipe for extra flavor and color, but the simple combination of lemons and salt always works a treat, and it can keep in your kitchen pantry for several months.

Wash and dry the lemons to remove any surface dust or dirt. Scrub the skins to remove any grit. Cut the ends off each lemon, then slice each fruit into quarters (if they are small) or eighths (if they are large). Remove the pith from the inside of each slice.

Put the lemon slices into a large bowl with the sea salt and toss well. If you are adding herbs or seasonings, add them now. Roughly speaking, you will need 2 or 3 teaspoons of salt for each slice of lemon.

Preparation time: 15 minutes
Makes: Approximately one 1-cup jar for every 2 or 3 lemons

Ingredients:
lemons
sea salt (about 8 teaspoons per lemon)

Variations:
For added flavor, try adding one of these seasonings
• whole cloves
• bay leaves
• crushed cinnamon sticks
• whole chili seeds

Push the lemon slices, one by one, into a sterilized jar. As you push, the juice should drain out and accumulate at the bottom of the jar. It is important that all the lemon pieces are completely immersed in fluid. If you find your jars are too full of liquid, drain a little from the top. You should not need to add any excess water as the juice will be squeezed out from each lemon.

When the jar is full, sprinkle the surface with salt, then seal the jar tightly. Leave it in a cool cupboard for 2 or 3 weeks before you start consuming the lemons. Each time you open the jar to use some of the lemon, sprinkle some salt on the surface when you're done. The lemons should last up to a year in your pantry.

preserved globe artichokes

Artichokes grow in abundance during hot European summers, and most traditional Mediterranean pantries would have included a few jars of these. They can cost a startling amount at the local deli, but if you preserve them yourself using your backyard harvest, you will pay only for the oil, the herbs, and the salt.

Peel the artichokes and cut them into quarters, removing the tough outer leaves. Place the vinegar, salt, and 4¼ cups of water in a small saucepan, then add the artichokes. Bring to a boil, then simmer for 10 minutes. Remove the artichokes from the saucepan and pat them dry. Let them drain for a couple of hours or overnight.

    Pack them into sterilized glass jars along with the herbs, then drizzle them with olive oil and tightly seal the jars. Let them sit for at least a week before consuming them. They will last several months in a cool pantry.

Preparation time: 5 minutes
Cooking time: 30 minutes
Makes: 1 large jar

Ingredients:
10 small globe artichokes
2 cups vinegar
sea salt
olive oil
fresh or dried herbs (oregano, thyme, or rosemary, or a combination)

QUICK & EASY TOMATO PRESERVE

Italian families often have a tomato bottling session as an annual ritual. They'll make enough to last the whole year, and no one goes home empty-handed. For those of us with just a few tomato plants, this quick and easy recipe will produce a couple of delicious jars of preserved tomatoes. They can be added to minestrone and other soups and stews and will last for several months in the freezer.

Preheat the oven to 300°F.

Put all the ingredients in a large ovenproof baking dish and stir well. Bake for 2 to 3 hours, or longer if you want a more concentrated sauce. Transfer the mixture to a deep mixing bowl and purée with a handheld blender. Place a large sieve over a second bowl and pour the sauce through a little at a time, pressing it through with a wooden spoon if you need to. Discard the solids left on the sieve and either bottle or freeze the liquid. If you are bottling and storing the preserve in the fridge, pour a few teaspoons of olive oil onto the surface of the preserve before sealing it. This acts as a lid and will prevent mold from forming.

Preparation time: 5 minutes
Cooking time: 3 hours
Makes: 2 cups

Ingredients:
4⅘ pounds ripe tomatoes, cored and cut into halves or quarters (depending on tomato size)
1 large onion, finely chopped
4 cloves garlic, crushed
⅓ cup olive oil
⅓ cup wine vinegar
½ teaspoon dried thyme
½ teaspoon dried rosemary
¼ cup fresh basil leaves

TOMATO KETCHUP

Okay—I know every supermarket stocks tomato sauce. This one, however, tastes far better than anything you will buy off the shelf. It contains no artificial colorings or preservatives and is an excellent way to use up over-ripe tomatoes during the summer months. When we buy large quantities of meat from our local farm, we usually get a few boxes of hand-made sausages. I devised this tomato ketchup recipe specially to complement the sausages, but it goes equally well with any meat dish, or as a dipping sauce with homemade hot potatoes.

Core and roughly chop the apples. Cut the tomatoes into halves. Peel the onions and cut them into quarters. Add all of these ingredients to a pot, cover them with water, and bring the water to a boil. Reduce the heat until the water is simmering, then add all of the remaining ingredients.

Simmer for 1½ to 2 hours on medium heat, until the apples are soft and well cooked. When the apples are soft, remove the pot from the heat and use a handheld blender or food processor to purée the mixture into a smooth paste.

Preparation time: 15 minutes
Cooking time: 12 hours or overnight
Makes: 1½ to 2 quarts

Ingredients:
3 medium apples
5 pounds tomatoes
2 medium onions
1 cup apple-cider or wine vinegar
¾ cup white wine
¼ teaspoon cayenne pepper
1 teaspoon ground cloves
1 teaspoon ground allspice
4 cloves garlic
¼ cup sea salt
1 teaspoon soy sauce
2 cups whole cane or brown sugar

Return the pot to a low heat. Let it gently simmer for 6 to 10 hours, or overnight, stirring occasionally to prevent it from sticking to the bottom. The mixture should thicken to the consistency of tomato sauce.

When it has thickened, pour the sauce into sterilized glass jars. Tip a few tablespoons of olive oil onto the surface of each jar before sealing it; this will prevent mold from forming. Store in a cool pantry or refrigerator.

pineapple Beer

Before the advent of sugar-laden soft drinks and fast foods, most households would regularly prepare their own home brew. Fermented soft drinks contain no artificial colors, flavors, or preservatives. The natural, wild fermentation process also develops healthy lactic-acid bacteria, which are good for the digestive tract.

The only drawback of natural fermentation is that the results can be a little unpredictable. Some ingredients ferment more quickly than others, depending on how much sugar they contain. You may find that your brew ferments more quickly than expected, or that it's extra bubbly or has extra bite!

To prepare your starter:

Day one: Combine all the ingredients in a small glass jar and stir well.

Days two through five: Add one teaspoon of ginger and one teaspoon of sugar each day, stirring well. After three days or so, you should start to see some bubbles and froth. If you don't, continue the process for another few days. The starter *must* be fed every day or it will die. Once it is

Preparation time: 10 minutes
Fermenting time: 3 days to establish the starter, 7 days for the beer
Makes: 4 quarts

Ingredients
FOR THE STARTER:
2 millilitres filtered water (if using tap water, boil it for 10 minutes, then return it to room temperature before using it)
2 teaspoons freshly ground ginger, skin and all
2 teaspoons white sugar
Plus additional fresh ginger and sugar. You'll need one teaspoon of each for every day it takes your starter to grow, so try to have a few teaspoons' worth of each on hand.

Ingredients continued ...

FOR THE BEER:
1 large ripe pineapple
   (approximately 2¼ pounds)
½ teaspoon sea salt
½ cup whole cane sugar
3 tablespoons freshly grated
   ginger
¾ tablespoon rice syrup
¾ cup whey

bubbling it is ready to use. If you don't see bubbles by day five, I would throw it out and start again (no hard feelings).

To prepare your beer:

Once your starter is bubbling and active, you are ready to go.

Coarsely chop the pineapple. In batches, purée it in a food processor. Gradually add the salt, sugar, and half the ginger, puréeing as you go to ensure that they are all well mixed in.

Pour the pineapple mixture into a large glass jug or bottle that has been sterilized, and measure the quantity as you go.

Add the starter and stir well. The final quantity of liquid should be about 3 quarts; so add some filtered water if you need to.

Cover the jug with a tea towel or muslin cloth and secure it with a rubber band. Place it in a warm place and allow it to ferment for 2 days, stirring once or twice each day so that the pulp doesn't just sit at the top.

After 2 days, strain the liquid to remove the pulp using a muslin cloth. Add 1 quart of water, the remaining ginger, and the rice syrup, and bottle it in sterilized, airtight bottles. The beer will be ready to drink in 2 days.

sauerkraut

Throughout history, in many parts of the world, cabbage has been preserved to create delectable fermented foods such as sauerkraut (Germany), choucroute (France), and kimchi (Korea, China and Japan). Archaeologists have found that even during the hunter-gatherer stage of human development, people fermented plants, and cabbage was a favorite.

A luscious and large cabbage, big enough to produce three quarts of sauerkraut, costs about four dollars at most organic stores. This quantity of sauerkraut will last you several months. I strongly recommended using a mandolin to shred the cabbage if you can get your hands on one. If not, slice it as finely as possible with a large, sharp knife (this can be a bit mind-numbing, but bear with it).

Finely shred or slice the cabbage and apples. Keep a couple of pieces of cabbage, preferably a couple of the large outer leaves, to one side.

Toss the shredded cabbage and apple together in a large bowl (in batches if there's too much to fit into the bowl at once). If you are using any extra herbs for flavor, add them now.

Preparation time: 20 minutes
Fermentation time: 7 days
Makes: about 4 quarts

Ingredients:
1 medium green cabbage (very fresh and ripe, and not going to seed)
1 green apple
2 tablespoons sea salt
¼ cup starter culture (optional—see page 307)

Variations:
For extra flavor, try adding dill, cumin, juniper berries, or caraway seeds.

With your bowl full of shredded cabbage, you're ready to start pounding. You need to pound the cabbage for a good 5 minutes, releasing the juice from the leaves. I use a traditional sauerkraut pounder that looks a bit like a large wooden hammer, but you can also use a potato masher or a large wooden spoon.

After you've pounded for 5 minutes or so, add the salt and stir well. If you are adding a starter culture, add it now. Toss well and pound for another minute or two. The salt should be evenly dispersed throughout the cabbage.

Pack the cabbage into sterilized jars. Fill each jar about four-fifths full (don't fill them completely, or the cabbage may spill when it ferments). As you are pressing the cabbage into the jars, juice will be released from the cabbage leaves. If there isn't enough liquid to completely cover the cabbage, you may need to add a few drops of filtered water mixed with salt (use ¾ tablespoon of salt for every 1 cup of water). You need to create an anaerobic environment (that is, an environment with no air) for the cabbage to ferment. As you add every piece of sauerkraut to the jar, check to be sure that the cabbage is tightly packed and there are no air pockets.

Place a cabbage leaf over the top of each jar as a protective lid, to shield the sauerkraut from the air. Then place a weight, such as a smaller glass jar or bottle filled with water, on top of the leaf. Alternatively, you can just apply a tight-fitting lid, but you will need to check it in a few days' time to ensure that the cabbage is still immersed in liquid.

It's ready to eat when it begins to taste bubbly and sour, usually in about 2 or 3 weeks. The exact fermentation period will depend on the weather and the local environment. Sauerkraut will ferment more quickly in warm temperatures than in cooler climates. Traditionally it was kept in a cool cellar and would last for months, sometimes years. I usually keep mine on our kitchen counter for the first 5 days, then transfer it to the refrigerator.

Resources

There is a wealth of information online for anyone interested in the fruga-vore approach to food. These sites below should help you to get started. And of course, you can visit me at www.frugavore.com.

Sourcing your food

Networking Association for Farm Direct Marketing and Agritourism (www.nafdma.com): A wonderful resource an direct producer-to-consumer agricultural relationships in the United States.

The Weston A. Price Foundation (www.westonaprice.org): The WAPF website provides information about nutrition and traditional cooking techniques, while local chapters can let you know what's happening nearby.

Slow Food (www.slowfood.com): Slow Food has branches all over the world. If you join your local convivium, you'll be kept up-to-date with news and events.

Real Milk (www.realmilk.com): A good place to start for information about whole milk, the philosophy behind it, and what's available in your area.

Gardening

Heirloom Organics (www.non-hybrid seeds.com): An online site that supplies information about non-hybrid seeds. You may also purchase your seeds through this site.

American Community Gardening Association (http://communitygarden .org) A wonderful resources for those intersted in finding community gardens.

Further afield

These additional sites are worth exploring, even from afar. Much of their advice is relevant wherever you are, and they provide an insight into what's happening elsewhere. Who knows: they might inspire you to start something similar in your neighborhood …

Local Harvest (USA) (www.localharvest.org): This site offers comprehensive information about sourcing organic food, growing heirloom vegetables, and getting involved in co-ops and community-supported agriculture. It also hosts a lively collection of food forums and blogs, so is a great place to ask questions and share ideas.

Locavores (USA) (www.locavores.com): If you're interested in learning more about the "locavore" movement, this is the place to start.

EatWild (USA) (www.eatwild.com): This site outlines the benefits—for humans and animals—of raising livestock on pasture.

The River Cottage (UK) (www.rivercottage.net): Through his *River Cottage* series of television programs, Hugh Fearnley-Whittingstall has become a champion of local, sustainable eating. The River Cottage website includes recipes and advice and recently launched a program to help put willing land-owners in touch with would-be community-gardeners.

The Ecologist (UK) (www.ecologist.org): Established in 1972 and now published online, *The Ecologist* was one of the first magazines to promote sustainable living. It provides practical tips and resources as well as news about events and campaigns.

Bibliography

Aubert, Claude (ed.). *Keeping Food Fresh: Old World Techniques and Recipes from the Gardeners and Farmers of Terre Vivante*. White River Junction, VT: Chelsea Green Publishing Company, 1999.

Blazey, Clive. *Growing Your Own Heirloom Vegetables*. Dromana: The Diggers Club, 2008.

Bornstein, S. & Bartov, I. "Studies on egg yolk pigmentation 1. A comparison between visual scoring of yolk color and colorimetric assay of yolk carotenoids," *Poultry Science*, 45 (2), 1996: 287–96.

Bourn, D. and Prescott, J. "A comparison of the nutritional value, sensory qualities, and food safety of organically and conventionally produced foods," *Critical Reviews in Food Science Nutrition*, 42(1), January 2002: 1–34.

Cordain, L., Eaton, S. B., Sebastian, A., Mann, N., Lindeberg, S., Watkins, B. A., O'Keefe, J. H., and Brand-Miller, J. "Origins and evolution of the Western diet: health implications for the 21st century," *American Journal of Clinical Nutrition*, 81, 2008: 341–54.

Daniel, Kaala. *The Whole Soy Story*. Winona Lake, IN: New Trends Publishing, 2005.

Davis, D. R., Epp, M. D. and Riordan, H. D. "Changes in USDA Food Composition Data for 43 Garden Crops, 1950–1999," *Journal of the American College of Nutrition*, 23(6), 2004: 669–82.

Enig, Mary. *Know Your Fats*. Silver Spring, MD: Bethesdapress, 2000.

Fallon, Sally and Enig, Mary. *Nourishing Traditions*. Winona Lake, IN: New Trends Publishing, 2000.

Fisher, Suze. "The quest for nutrient-dense food: high-brix farming and gardening," *Wise Traditions*, March 17, 2005.

French, Jackie. *Jackie French's Chicken Book*. Melbourne: Aird Books, 1993.

Halweil, Brian. "Still no free lunch: nutrient levels in US food supply eroded by pursuit of high yields," *Critical Issue Report*. Foster, RI: Organic Center, 2007.

Hui, Y. H. (ed.) *Handbook of Food Science, Technology and Engineering, Volume 2*. Boca Raton, FL: Taylor and Francis, 2006.

Katz, Sandor Ellix. *Wild Fermentation*. White River, VT: Chelsea Green Publishing, 2003.

Leenhardt, F. et al. "Moderate decrease of pH by sourdough fermentation is sufficient to reduce phytate content of whole wheat flour through endogenous phytase activity," *Journal of Agriculture and Food Chemistry*, 53(1), 2005: 98–102.

Luard, Elisabeth. *European Peasant Cookery*. London: Grub Street Press, 2004.

Mayer, Anne-Marie. "Historical changes in the mineral content of fruits and vegetables," *British Food Journal*, 99(6), 1997: 207–11.

McGee, Harold. *On Food and Cooking: The Science and Lore of the Kitchen*. New York, NY: Simon and Schuster, 2004.

Mozaffarian, D., Katan, M. B., Ascherio, A., Stampfer, M. J. and Willett, W. C. "Trans-fatty acids and cardiovascular disease," *The New England Journal of Medicine*, 354, 2006: 1601–13.

McLagan, Jennifer. *Fat: An Appreciation of a Misunderstood Ingredient, with Recipes*. New York, NY: Ten Speed Press, 2008.

Nagel, Ramiel. "Living with phytic acid," *Wise Traditions*, March 2010.

O'Keefe, S. F. "An overview of oils and fats, with a special emphasis on olive oil," In Kiple, K. F. and Ornelas, K. C. (eds), *The Cambridge World History of Food, Volume One.* Cambridge: Cambridge University Press, 2000.

Pollan, Michael. *In Defence of Food.* New York, NY: Penguin Books, 2008.

Pollard, Tessa. *Western Diseases: An Evolutionary Perspective.* Cambridge: Cambridge University Press, 2008.

Price, Weston. *Nutrition and Physical Degeneration.* New York, NY: Keats Publishing, first published in 1939; sixth edition published in 2003.

Reganold, J. P. et al. "Sustainability of three apple production systems," *Nature,* 410 (April 19, 2001): 926-30.

Robinson, Jo. *Pasture Perfect.* Vashon Island, WA: Vashon Island Press, 2004.

Schmid, Ron. *The Untold Story of Milk.* Winona Lake, IN: New Trends Publishing, 2003.

Schroeder, H. A. "Losses of vitamins and trace minerals resulting from processing and preservation of foods," *American Journal of Clinical Nutrition,* May 24, 1971: 562–73.

Shurtleff, W. and Aoyagi, A. "A special report on the history of soy oil, soybean meal and modern soy protein products," Lafayette, CA: Soyfoods Center, 2004.

Siri-Tarino, P. W., Sun, Q., Hu, F. B., and Krauss, R. M. "Meta-analysis of prospective cohort studies evaluating the association of saturated fat with cardiovascular disease," *American Journal of Clinical Nutrition,* 91(3), March 2010: 535–46.

Siri-Tarino, P. W., Sun, Q., Hu, F. B. and Krauss, R. M. "Saturated fat, carbohydrate, and cardiovascular disease," *American Journal of Clinical Nutrition,* 91(3), March 2010: 502–9.

Stewart, Robin. *Robin Stewart's Allergy-free Home.* Melbourne: Black Inc., 2002.

Temple, N. J. and Burkitt, D. P. *Western Diseases: Their Dietary Prevention and Reversibility.* Totowa, NJ: Humana Press, 1994.

The Australia Institute. *Wasteful Consumption in Australia.* Discussion paper number 77, March 2005.

Thomas, David. "A study on the mineral depletion of the foods available to us as a nation over the period 1940 to 1991," *Nutrition and Health,* 17(2), 2003: 85–115.

Trubek, Amy. *The Taste of Place: A Cultural Journey Into Terroir.* Berkeley, CA: University of California Press, 2008.

Willett, W. C. and Ascherio, A. "Trans-fatty acids: are the effects only marginal?" *American Journal of Public Health,* 84(5), 1994: 722–24.

Williams, C. M. "Nutritional quality of organic food: shades of grey or shades of green?', *Proceedings of the Nutrition Society,* 61(1), February 2002: 19–24.

Acknowledgments

This book would not have been possible without the sound encouragement of so many people along the way. Thank you foremost to Denise O'Dea at Black Inc., a kind and helpful ear and lover of my orange marmalade.

Thank you to the many experts who have lent an ear or a bottle of milk along the way: Ron and Beverly Smith, thank you for taking me in, showing me your farm, and teaching me all about "salad-bar" grazing. Thank you to Sally Fallon, Patricia O'Donnell, Marie Danvers, and Ron Hull for your feedback and valuable input.

Thank you to Michael, Pat, and Giules—this book was created with you in mind and would never have been possible without your kind enthusiasm for my cooking and your support of all things mad and wonderful at Drummond Street.

Thank you to my dear family: J, Bec, Mum, and Alistair for teaching me to cook and fostering the growth of several chickens and other wild animals in our expansive and eccentric household.

Finally, and not at all frugally, thank you to my partner, milkman, and fellow gardener Darryl. Thank you for all of our trips to the op-shops, the farms, and the nurseries, and for reading through countless drafts. It's been great fun.

Index of Recipes